Cognition and Tool Use

Cognition and Tool Use
Forms of engagement in human and animal use of tools

Christopher Baber

CRC Press
Taylor & Francis Group
Boca Raton London New York

CRC Press is an imprint of the
Taylor & Francis Group, an **informa** business

A TAYLOR & FRANCIS BOOK

CRC Press
Taylor & Francis Group
6000 Broken Sound Parkway NW, Suite 300
Boca Raton, FL 33487-2742

First issued in paperback 2020

ISBN-13: 978-0-367-45447-0 (pbk)
ISBN-13: 978-0-415-27728-0 (hbk)

Visit the Taylor & Francis Web site at
http://www.taylorandfrancis.com

and the CRC Press Web site at
http://www.crcpress.com

Typeset in Sabon by
Newgen Imaging Systems (P) Ltd, Chennai, India

British Library Cataloguing in Publication Data
A catalogue record for this book is available from the British Library

Library of Congress Cataloging in Publication Data
Baber, Christopher, 1964–
    Cognition and tool use: forms of engagement in human and animal use of tools /
Christopher Baber.
        p. cm.
    Includes bibliographical references and index.
        1. Cognition and culture. 2. Tools. 3. Tool use in animals. I. Title.
BF311.B228 2003
152.3'8—dc21   2003040250

And you may ask yourself,
how do I work this?

Talking Heads, *Once in a Lifetime*

For
Sara Megan Liam Natasha Jessica

Contents

Figures

Tables

Acknowledgements

This book has been in preparation for several years and began in discussions with Ted Megaw. As I worked on the ideas contained in the book, I was able to give presentations to the Educational Technology and the Sensory Motor Neuroscience Research Groups at Birmingham University. Various drafts of various chapters have been read by Stamatina Anastopoulou, Anthony Costello, James Knight and Alan Wing. My thanks to all of these people for their comments, insights and criticisms. I hope that this book does justice to their input.

1   Introduction

Neither the naked hand nor the understanding left to itself can effect much. It is by instruments and aids that work is done.

(Francis Bacon, 1620)

Introduction

This book is about the uses to which people and animals put tools. The main focus of the book is Ergonomics, i.e., the 'laws' (nomos) that govern or relate to 'work' (ergo). When I use the term 'ergonomics', I mean the relationship between animate beings and their environment. Thus, whenever some aspect of the environment, or objects in that environment, impinges on the activity of a person (or animal), one can see a role for ergonomics. For the most part, the scientific study of ergonomics has been concerned with studying the effects of the environment on human performance, and then seeking to modify or otherwise redesign the environment. Redesign might focus, for instance, on changing the environment to better suit human stature, to better control thermal comfort, or to support more effective and efficient performance. Redesign might also focus on changing objects within the environment, which is the concern of this book. It is my contention that a clear and detailed understanding of the manner in which people make use of tools can offer a significant contribution to the study of ergonomics. Of course, the study of how people make use of tools is not the sole province of ergonomics, but is also important to anthropology, archaeology and a host of other human sciences. Having said this, there has been surprisingly little attempt to bring together all the different disciplines that have a bearing on this topic.

When I use the term 'tool', I have in mind physical implements and artefacts that are used to make changes to other objects in the environment. The physicality of tools is important for this book because one of the key issues that I am exploring is the relationship between physical and cognitive activity. A commonly used example of tool use is the act of hammering a nail (Figure 1.1).

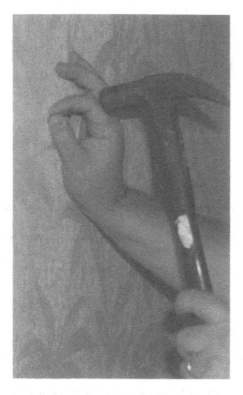

Figure 1.1   Hammering a nail into a wall.

This is a mundane enough activity, and readers will be forgiven if they struggle to see a cognitive component to what is clearly a physical action of swinging a hammer onto a nail. There are four points that can be presented to demonstrate the central argument of this book. First, in order to strike the nail's head cleanly and sharply, one needs to be able to define an appropriate trajectory and force in which to move the hammer. If one watches very young children playing with peg-and-hammer sets, one realizes that such coordination is not innate, but develops (through experience, practice and maturation). It is proposed that such definition requires cognitive activity (rather than merely muscular coordination). Second, if a hammer is not available, then (adult) humans are able to adapt other objects, such as the heel of a shoe. This suggests that tools have several 'meanings', e.g., the tool itself, the function that the tool can perform and the potential role of an object in achieving the users' goal. If this is the case, then the manner in which humans (and animals) learn to represent tools and their functions could provide interesting insight into how tools are used. Third, the smooth performance of hammering proceeds in such a manner as to make one 'forget' the hammer and focus on the goal of driving home the nail. It is only when the action 'breaks down', e.g., one misses the nail or the nail hits a knot in the wood or the nail does not appear straight, that one refocuses one's attention on the hammer in one's hand. This, for me, is an interesting aspect of using tools, i.e., through practice, the tool 'disappears' from one's immediate awareness by becoming both part of the person and part of the task. Fourth, the manner in which even the simplest tools, like hammers, are wielded by the novice (or at least intermittent) hobbyist and the expert craftsperson testify to distinguishable levels of skill that reflect both a tradition of working and a learned relationship between tool and user.

Tools mediate our engagement with the world. Thus, hammering modifies the length of the nail, the contact between the nail and whatever it is being driven into, the goal of joining two pieces of wood (or fixing a hanging for a picture, or repairing a drawer, etc.), the impact of hitting the hammer felt on the wrist via the handle etc. The actions are on the tools themselves, on the objects that the tools affect, on the users of the tools, on the surrounding environment, on our perception of the world and the goals that we hold. In one form, this means that we need to consider the basic mechanics and physical performance involved in using tools. In another form, using a tool requires some elements of planning and coordination of motor activities that have a basis in cognition, and some form of representation of tools and their functions. In a further form this requires us to concentrate on aspects of everyday cognition, i.e., the issues relating to the 'feel' of the tool and the ways in which tools 'disappear' in use. My proposal is that each manner in which tools mediate our engagement with the world represents a specific set of knowledge and perception–action coupling, and that the integration of these forms underlies tool use.

In order to think about using tools it becomes necessary to weave together a story that combines elements of experimental psychology, sociology, anthropology and ergonomics, with contributions from other disciplines. Of course, such a cast will inevitably lead to accusations of misrepresentation or favouritism from exponents of the various disciplines that are called upon. However, I propose that tool use is sufficiently complex to warrant the efforts of many researchers, looking at the problems in many forms and proposing many answers; my hope is to pull together some of these strands into a coherent, interesting and informative account.

From this perspective, research into early hominid and animal activities relating to tools can be used to highlight differences between these species and modern humans. I realize that such a position has not been particularly fashionable for the past 100 years or so; most writers on tools emphasize that, because tool use can be found across many species, it is no longer the sole province of humanity. Thus, the notion that 'man is a tool using animal' [1] is no longer seen as one of the distinguishing features of humankind. By implication, writers suggest that the distinction between humans and animals, in terms of tool use, are slight – proving once again that humans are merely animals. While I do not have a quibble with the tenor of this argument, my feeling is that human tool use *does* differ, and differ significantly, from animal tool use, and that such differences point to essential and elemental differences between humans and animals. Therefore, when discussing how animals use tools, my focus will be on contrasting these activities with human tool use. Furthermore, it is probably a good time to point out that I see no sense in linking tool use with 'intelligence' (although this has been a particular hot potato amongst many commentators on tool use, particularly from animal observations and anthropology). It seems to me that tool use as a way of mediating one's activity on the world will be as complex as necessary; thus, to speak of an ant carrying food in a leaf as being less intelligent than a human carrying food on a plate seems to imply dimensions of intelligence that allow direct comparison between these species. The form of engagement that is appropriate to the tool user, in order to effect seamless mediation of activity and to achieve optimum performance, would seem to me to be the only necessary criteria when considering tool use – and to speak of some aspects of these forms of engagement as being more 'intelligent' than others seems to miss the point.

What is a tool?

There are two basic definitions of the word 'tool'. The first refers to any handheld implement that can be used to perform a task, such as a hammer, a knife or a fork. The second refers to any form of support that can be drawn on to help perform a task. In this second definition, it is possible to think of a tool as a piece of what McCullough [2] calls applied intelligence, i.e., technology that allows us to expand upon the limited repertoire of manual and cognitive skills that we possess. My feeling is that it is this expansion of skills we possess that sets tools apart from other machinery and technology; a tool allows us to take an existing operation and, in some sense, amplify it. Both definitions of the word 'tool' share a notion of objects in the world that help us to do things.

As mentioned above, I am assuming that a tool is a physical object. This limits the range of items that can be considered, and already introduces an element of controversy. One of the most prescient writers on the subject of human tool-use, Vygotsky [3] observed that a defining feature of human behaviour was the ability to internalize tools. By this he meant that children develop the ability to use 'psychological tools' to help perform cognitive tasks. In other words, one could use a simple 'tool', such as beads, when learning to count, but then dispense with the physical artefacts when a level of proficiency has been attained. From this level of proficiency, i.e., when one can count without the aid of physical objects, one can also perform manipulations on the internalized representation, e.g., one can multiply, add, divide or subtract numbers without using physical objects. However, once the manipulations become too complex or detailed to perform mentally, then one again makes use of physical representation; in this case, one uses written symbols to record the steps in the calculations. This indicates that in describing tools merely as physical objects, I am walking a rather tricky path between different levels of representation. When one speaks of 'internalizing' tools, the idea of 'amplification' is clearly inappropriate; rather what one is doing is representing aspects of the world in new ways and manipulating these representations. This is taking us some distance from the focus on tools as physical objects.

To make matters worse, the word tool has been, to use a phrase from linguistics, 'over-extended'; in other words, the definition of tool has been stretched to such an extent that it means more or less anything that a person can use to help to do some task. Thus, pencil and paper could constitute a tool, as could the numbers written on the paper, as could one's fingers when one is doing maths. Indeed, in the word processing package in which this chapter is being written, there is a menu called 'Tools', which contains such diverse functions as a spell-checker, mail merge and something called 'options'. From this discussion, one can see that the word 'tool' usually means any form of crutch or support to assist the person in doing something, regardless of what form the support might take.

I will consider the role of 'cognitive artefacts' in human performance in Chapter 10. Cognitive artefacts are those representations, devices and objects that people use to support cognitive performance. For example, a shopping list supports both the activity of remembering what to buy and also the activity of planning what to buy. Alternatively, a calculator supports the activities of manipulating numbers, recording the outcome of those manipulations and performing manipulations that are too time-consuming or complex to perform manually or mentally. I hope that by the time we turn to cognitive artefacts, we would have acquired sufficient insight into the use of physical tools to be able to consider how and why such objects are useful. The consideration of cognitive artefacts will also assist in developing the arguments relating to representation because these objects are essentially external representations in support of cognitive activities. My proposal is that, perhaps paradoxically, physical tools also function as external representations in support of cognitive activity.

Having considered some of the ways in which 'tool' has been defined beyond its scope, it is now worth reviewing the various definitions that other researchers have proposed. Ostensibly, the word 'tool' is derived from the Old English *Tawian*, meaning to prepare for use. While this tells us little about the word 'tool' itself, the definition does imply that tools are used for specific purposes. Indeed, Samuel Butler wrote that 'Strictly speaking, nothing is a tool unless during use' [4]. In other words, a tool is something that acquires significance through the functions that it can be used to perform or support. To a certain extent this proposal echoes the distinction that Heidegger [5] drew between things we think about and things which we use, i.e., between those objects which are *vorhanden* and those which are *zuhanden*. In this distinction, *vorhanden* refers to the theoretical aspects of objects which allow people to contemplate these objects. I suppose if the reader was asked to 'describe a screwdriver', it is possible to produce a verbal description containing handle, shaft and head. On the other hand, *zuhanden* refers to the user's relationship to objects in use. This notion of *zuhanden* seems to reflect Butler's proposal that 'nothing is a tool unless during use'. In other words, the meaning of the screwdriver comes as much from the manner in which it is used, as in any verbal description that one can apply to it. Indeed, one could even say that a verbal description is a very poor means of describing the screwdriver; it would be better to offer a combination of words, pictures and mimed actions (or better still to show a screwdriver being used).

This distinction between the *vorhanden* and *zuhanden* of tools implies two distinct types of knowledge (although it is debatable as to whether *zuhanden* constitutes 'knowledge' in any sense in which facts can be represented and discussed). What is apparent is that, through the use of a tool (and any object that one physically manipulates) the focus of attention can be less of the object and more on the task; in other words, the object 'disappears' during use and one is aware of the performance of a task rather than the feeling of the object in one's hands. This is true only in so far as the task proceeds without problem. As soon as a problem occurs, the attention shifts from the performance of the task to the presence of the object. The implication is that tool use becomes a dance between *vorhanden* and *zuhanden*, between a practical, physical engagement between person and task and a theoretical understanding of the properties of the object. To develop this into an example, imagine using a file to shape a piece of metal (Figure 1.2). As you move the file back and forth across the metal, you become caught up in the rhythm of filing and focus on the angle and movement of the file. To check the result, you need to stop and, perhaps, inspect the object being worked. If the file slips or sticks, then you might look from the object to the file's edge or to the way in which your hand is grasping the handle or to the angle that the file is moving across the metal etc. In the first part of the activity, the focus of attention is on the action of filing the object, but in the second part of the activity, the focus of attention is on the file itself.

Butler goes on to state, 'The essence of a tool, therefore, lies in something outside the tool itself. It is not in the head of the hammer, nor in the handle, nor in the combination of the two that the essence of mechanical characteristics exists, but in the recognition of its unity and in the force directed through it in virtue of this recognition' [6]. My interpretation of this quotation is that, in order to effectively use a tool, one must be able to interpret (and represent) the functions that the tool can support and the actions that one needs to perform in order to elicit these functions. In this manner, people can, 'see' a shoe, a rock or a piece of wood as a hammer. The point of this discussion is that I believe that tool use has a strong cognitive

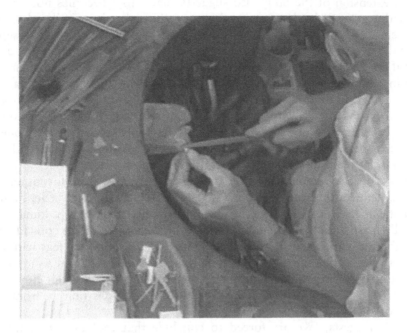

Figure 1.2   A jeweller filing a ring.

component that can handle these different forms of representation, and that a cognitive perspective that views knowledge as essentially declarative (i.e., something that one does with words) is unable to cope with tools. This might be one reason why so little has been written about tools from the perspective of cognitive psychology. Having said this, recent theorists have been exploring the influence of context on cognition, emphasizing the interaction between thinking and activity, and the role that artefacts play in this interaction. For the purpose of this book, such approaches represent a way of considering the *zuhanden* of tools.

From the literature of animal tool-use, there have been several attempts at describing what is meant by a 'tool'. For instance, Goodall, who has spent many years studying the behaviour of chimpanzees, proposes that 'tools' involve '… the use of an external object as a functional extension of mouth or beak, hand or claw, in the attainment of an immediate goal' [7]. This definition highlights two important features. First, like many writers, Goodall emphasizes the notion that 'tools' represent an extension of the body. As Beck, in his comprehensive review of animal tool-use observes, this means that, for animals, a tool will '… extend the user's reach, amplify the mechanical force that the user can exert on the environment, enhance the effectiveness of the user's display behaviours, or increase the efficiency with which the user can control fluids' [8]. However, the notion that a tool is an extension of the body also suggests that a tool becomes not merely something that is held or carried, but something that is assimilated to the movement of the body to such an extent that it becomes difficult to separate the two. Obviously, one can visually distinguish the tool from the user, but if one considers the experience of using well-made cutlery to eat a meal, for instance, one is less aware of manipulating the knife or fork than of preparing food and moving it to one's mouth. Of course, should the tools lead to problems, e.g., if the knife is not able to cut a particularly tough piece of meat, or if one cannot manipulate the tools properly, e.g., if one is an inexperienced user of chopsticks, then the assimilation is broken and one is aware of the tools as objects separate from one's body.

Goodall also emphasizes the fact that tools are used for attaining goals. Thus, a dog running along a field carrying a branch in its mouth might not be using a tool (as it does not appear to have any immediate 'goal', apart from the pleasure of biting onto something or the coincidental delight it might obtain from thwacking its owner around the legs with the leaves), but a chimpanzee using a twig to extract termites from a mound is using a tool. This raises some interesting considerations about how one defines 'goal' for tool use, and also how one excludes objects from the category 'tool'.

Beck suggests, 'We are forced to conclude that tool use, in terms of topography, function or causal dynamics, dovetails imperceptibly with other categories of behaviour' [9]. One feels that Beck is somewhat disappointed by this 'dovetailing', in that it makes the scientific study of tool use *per se* difficult, particularly when this study involves field-based observation of animals performing activities that may or may not constitute the use of tools. For the purposes of this book, the quotation is interesting in that 'dovetailing' of tool use with other behaviours is an essential component of the argument being developed. In order to understand and appreciate how tools are used, one needs to be able to consider a broad range of behaviours. By extension to this proposal, when one modifies a 'tool', e.g., through redesign, one is also effecting changes in other aspects of behaviour. For example, to return to the example of hammering a nail: consider how a user's actions might change if the head of the hammer was either small and light (such as a tack-hammer) or heavy and large (such as a claw-hammer). Obviously, there are mechanical changes, relating to weight, movement, etc., but there might also be other changes, e.g., given a heavy hammer the user might attempt smaller swings and expect to use few impacts to drive home a small nail on which to hang a picture.

To conclude, a tool is a physical object that is manipulated by users in such a manner as to both affect change in some aspect of the environment and also to represent an extension of the users themselves. The manipulation is directed towards a specific goal or purpose, and the associated activity requires a degree of control and coordination. From an ergonomics perspective, a set of criteria for tool design was proposed in the 1960s by Drillis [10]:

1 The tool must be able to perform as well as possible the function for which it was intended.
2 The tool must be properly proportioned to both the dimensions of the user and the constraints of the task.
3 The tool should be designed to optimize the performance of the user, in terms of the users' strength, work capacity, etc.
4 The tool should be designed to minimize costs in use, in terms of fatigue, injury, etc.
5 The tool must provide appropriate and useful sensory feedback to the user, or at least, use of the tool should provide an opportunity for the user to monitor performance and effect appropriate changes, i.e., the use of tools implies perception–action coupling through which the characteristics of the tools constrain and influence the user's movements.

To this list, I would add a sixth criterion:

6 A tool extends an existing ability, by increasing the power or precision with which an action can be performed.

While these criteria were developed for human tool-use, they can apply equally well to animal tool-use. For example, Table 1.1 compares the general activity of 'hammering' between human and chimpanzee, in terms of these criteria.

Ostensibly, Table 1.1 demonstrates that Drillis's criteria can be applied to chimpanzee tool-use. More subtly, however, these criteria can be used to

Table 1.1 Comparing human and chimpanzee tool-use against ergonomic criteria

Design criteria	Human	Chimpanzee
Perform intended function	Drive nail into wood	Crack open a nut
Properly portioned	Handle should fit hand, head should not be too heavy	Stone should fit hand, and should be correct weight
Optimize performance	Translate kinetic energy to impact force	Translate kinetic energy to impact force
Minimize risk	Strike nail head, miss fingers	Hold nut firmly on anvil
Feedback	Impact force, movement of nail, reducing length of nail	Nut cracks, nut opens
Extend existing ability	Bang hand onto object	Hold object in hand and bang against surface

define the choice of stone by the chimpanzee. As we shall see in Chapter 3, nut-cracking is not a straightforward process but involves selection of a good anvil stone and selection of a hammer stone that is both heavy enough to crack the nut and light enough to be wielded by the chimpanzee and to prevent the nut from shattering. In this example, tool use becomes as much a matter of 'creating' the correct tool as using the tool. Chimpanzees (and humans) are able to use a variety of objects to perform similar functions, e.g., stones, sticks, shoes can all function as hammers. This raises questions about the ways in which objects 'become' tools.

Tools as 'augmentation means'

It has already been noted that tools extend the range of activities that can be performed, e.g., by extending reach, amplifying force or otherwise changing the manner in which the user can change objects in the environment. Tools also modify performance itself, e.g., consider eating with chopsticks or with a knife and fork. Engelbart [11] proposed four characteristics by which 'augmentation means' could influence performance:

1   Tools and artefacts: the technologies that we use to work on the world which supplement, complement or extend our physical and/or cognitive abilities.
2   Praxis: the accumulation and exploitation of skills relating to purposeful behaviour in both work and everyday activity.
3   Language: the manipulation and communication of concepts.
4   Adaptation: the manner in which people could (or should) adapt their physical and cognitive activity to accommodate the demands of technology.

This set of characteristics of augmentation means can apply to all aspects of technology, but for the purposes of this book, raise some key questions about our relationship with tools. The first characteristics, 'tools and artefacts', have already been considered to some extent in the previous discussion about what constitutes tools. The second characteristic, 'praxis', introduces the notion that when we use tools we change our manner of working and that, by implication, different tools evoke different changes in work activity. Thus, e.g., using a shoe to bang in a nail leads to definable changes in activity in comparison with a hammer. My proposal is that these changes in work activity are as much cognitive as physical. The third characteristic, 'language', calls to mind two points. The first is simply the manner in which tool-using skills are communicated between, say parent and adult. Chapter 3 explores the acquisition and transmission of tool-using skills in humans and animals (particularly chimpanzees and other apes). In this manner, 'language' (as the ability to communicate) underlies the acquisition and sharing of skills relating to tool use. The second, and I feel more interesting point, relates to the notion that tools are themselves information bearers. Returning to the point raised earlier, that tools are items of 'applied intelligence', one can say that the design of a tool represents the crystallization of a specific notion of how one should perform a task. Thus, there are many designs of hammer or shovels, and each design (subtly different from its counterparts) reflects a particular way of performing a task, e.g., a heavier head or a longer handle reflect different notions of how one ought to swing the hammer. In this manner, I assume that tools are representations of specific views of working and that the 'language' of tools *per se* provides an interesting and enlightening avenue of exploration. This point will be developed further in Chapter 7. Finally, tools require some modification to activity. Given that tools, typically, represent extensions of the user's limbs, then one level of adaptation is simply the question of learning to manipulate an extended limb. However, the ability to skilfully use a tool also requires the coordination of tool and limb in a manner to produce optimal performance with minimal effort.

Everyday cognition

While the word 'cognition' implies a particular approach to the scientific study of human behaviour, there is growing debate regarding how one studies the various processes which make up cognition. In broad terms, cognition relates the ways in which people (and animals) draw information from the world, combine this with the knowledge that they already possess and interpret or make decisions about the information. Thus, one view of cognition is that it is a form of information processing. Indeed, this view relates to the popular notion of the brain as a 'computer' that can process information according to programmes and routines. Like many popular notions, such a view is only partly true and a great deal of research has addressed the differences between the brain and computers. This is not the place to review this work, but it is worth noting that human information processing seldom runs according to strictly defined, pre-programmed routines. Furthermore, it is likely that human information processing is both hierarchically organized, with different information being processed in more detail and with greater effort than others, and functionally separated, with specific areas of the brain addressing specific types of information.

Much of the research into human cognition has tended to be laboratory based. This provides a means of conducting well-controlled experiments that allow focus on specific aspects of information processing. However, there is a parallel trend, going back to Neisser's call for 'ecological validity' in the study of human behaviour, arguing that cognition should be studied in everyday settings. This call reminds me of the research into decision-making that demonstrated that people rarely perform in the manner that rational, logical, predictive models of decision-making would suggest.

Indeed, when one considers human performance in everyday settings it appears to exhibit similarities to the behaviour studied in the laboratory but also some striking differences. For instance, Lave [12] studied people performing supermarket shopping tasks. Many of the tasks involved solving mathematical problems, such as 'does product X with 25 per cent extra free represent better value than product Y?' or 'if I buy product X will this be sufficient for eight people coming to dinner tomorrow?' When these problems were expressed using the formula and terminology of mathematics, very few of the participants were able to solve the problems, but they could all solve the problems in the supermarket. At one level, this indicates that the manner in which a problem is presented modifies the approach that one can take to solve it. At another level, it implies that we have all manner of strategies that 'work' in everyday life and that can be applied with little or no effort in order to solve problems.

In broader terms, one can consider that these strategies might vary across cultures, e.g., in relation to working practices for different industries or in relation to different countries. For example, a commonly used example is the difference in operation between UK and US light switches, i.e., do you press the switch up or down to turn on the light? This relates to the notion of population stereotypes. Ergonomics has been, since the 1950s, concerned with how people 'know' whether you need to turn a radio volume control clockwise or anticlockwise to increase the volume. The general consensus seems to be that we have well-defined stereotypical responses that influence our interactions with simple products. There is some evidence to suggest that these responses vary between countries, and that they can be influenced by the design of the product that we are using. For example, people will employ a 'turn clockwise to increase' stereotype, unless the control knob is placed beneath the slider they are controlling. The notion of population stereotypes implies that people possess schema of appropriate responses to controls, and has echoes of the notion of affordance that has already been discussed in this book. This idea will be developed further in Chapter 11.

It seems to me that tool using is an area that could be considered under the general heading of everyday cognition. We rarely focus attention on the use of a tool, particularly when that tool is highly familiar. I will be using the example of eating with knives, forks and spoons as a running example in this book; eating is so routine an activity for most of us that we do not need to focus particular attention on the use of cutlery. However, the fact that we possess this 'skill' suggests that we should treat it with the same respect we accord to other skills.

Forms of engagement

In this chapter, I have defined a 'tool' as a physical object that is manipulated by users in such a manner as to both affect change in some aspect of the environment and also to represent an extension of the users themselves. The manipulation is directed towards a specific goal or purpose, and the associated activity requires a degree of control and coordination. From this definition, I view tools as objects external to the user that support engagement with the world around the user. The central thesis of this book is that there are several forms of this engagement. By considering how different examples of tool use relate to these forms of engagement, I feel that we can come to a better understanding of how tools are used. When I use the term 'engagement', I have in mind both the notion of interaction that is often used when speaking of using technology (i.e., as in human–computer interaction), and the notion of physically engaging with an object (i.e., as in gears engaging), and also in the sense of being involved with something to the extent that it becomes the focus of attention (i.e., in the sense of being engrossed in a film). I realize that this will tend to make 'engagement' a portmanteau word (and one that is rather untidily packed). However, this provides some indication of the multiple dimensions of activity that I feel are relevant to an appreciation of tool use.

I propose six forms of engagement to describe tool use: environmental, morphological, motor, perceptual, cognitive and cultural. While the following section provides short definitions of each term, it is important to realize that these terms are being developed through the coming chapters. In other words, the terms, at present, provide a framework for thinking about how animals and people use tools and are used to develop a theory of tool use.

By environmental engagement, I mean the ability of an organism to respond to aspects of the environment. Such responses could be innate, such that the presence of a particular feature in the environment will evoke a specific action, or they could be learned, e.g., through stimulus–response (S–R) conditioning, or they could represent particular perception–action couplings.

By morphological engagement I mean the ability of an organism to use hands, claws, mouth, beak, mandibles, etc. to grasp and wield objects. The dimensions of the object will relate to the morphology of the organism holding it.

By motor engagement, I mean the ability to manipulate objects. This relates to morphological engagement, in that the type of hold will be affected by the organism's morphology. However, sophisticated motor engagement might involve the organism exhibiting a variety of grips and changing the grips depending on the task at hand. Thus, motor engagement reflects both the postures adopted during the use of tools and also the control and coordination of movement.

By perceptual engagement I mean that ability to interpret feedback from using the tool, and relate this feedback to a particular set of expectations.

By cognitive engagement, I mean the ability to represent the function of tools and to represent the characteristics of tools, as well as the ability to coordinate actions through psychomotor skills, and the ability to relate tools to goals.

By cultural engagement I mean the ability of the organism to acquire tool-using skills from other animals (as opposed to being born with the ability), and the way in which tool use reflects certain traditions of action. Thus, a jigsaw, for instance, is designed to support a particular type of sawing activity which not only involves certain patterns of motor activity, i.e., the 'correct' way to use the saw, but also reflects certain desirable consequences of this activity, i.e., the need to produce decorative shapes from wood.

Taken together, it is proposed that consideration of tool use requires a multidisciplinary perspective. I have taken the notion of forms of engagement as my starting point for this book. By asking how, for instance, cultural engagement and morphological engagement might interact in specific examples of tool use, such as in chimpanzees using twigs to fish for termites or in farriers using hammers to make horseshoes, we can develop a richer and more detailed picture of the role of tools in our lives. Furthermore, this multidisciplinary approach can skip over some of the debates relating to tool use, e.g., does tool use reflect intelligence? Are humans 'better' at using tools than animals? etc.; my response to these questions is that they are at best misleading and that the manner in which a tool is used reflects the appropriate forms of engagement for that tool being used by that species in that environment in order to achieve that goal.

The structure of the book

Having defined what I mean when I use the word 'tool' and introduced the notion of forms of engagement in this chapter, the remaining eleven chapters explore some of the implications of these ideas. In Chapter 2, reports of tool use by insects, fish, birds and mammals are examined. The main focus is on tool use in the wild, with a view to appreciating the sort of tool-using activities that these species exhibit. In Chapter 3, we turn our attention to tool use by primates, particularly chimpanzees. This activity will be contrasted with some aspects of object manipulation by young children. In Chapter 4, we consider the manner in which tools are made, particularly to develop further the contrast between human and animal tool-use. Chapter 5 focuses on the manner in which humans use tools in their work, with particular reference to skilled tool users. Following this Chapters 6 and 7 look at two aspects of the design of tools: Chapter 6 is concerned with the general ergonomics of tool design, and Chapter 7 focuses on the ways in which people ascribe different meanings to tools. Chapter 8 considers the ways in which tool use can break down, particularly in terms of human error and neurological impairment. Chapter 9 develops the notion of cognitive artefacts, discussed earlier in this chapter, and Chapter 10 explores the developments in tool design for the coming century. Following the discussions of tool use in the previous chapters, Chapter 11 sets out an initial theory of tool use, combining forms of engagement with the notion of motor and cognitive schema. The book ends with a chapter on conclusions that examines the implications for the notion of forms of engagement and tool use.

2   How animals use tools

But what from the very first distinguishes the most incompetent architect
from the best of bees, is that the architect has built a cell in his head before
he constructs it in wax.

(Marx, 1890)

Introduction

In *The Origin of Species*, Charles Darwin wrote of animals using tools.
For instance, monkeys using stone hammers to crack open nuts, and
elephants using branches to wave away flies. He speculated that this behaviour was the result of the animals compensating for biological shortcom-
ings, e.g., teeth that were too weak to crack the nuts or trunks too short to
reach tails. One implication of this view is that a 'tool' serves as an extension of the animal and is selected to extend behaviours that the animal
already practiced, rather than representing novel sets of behaviours.

A key point to note from Darwin's observations is that, despite being
reported in 1871, very little work was done on animal tool-use for a further
century until Jane Goodall's pictures of chimpanzees using tools (published
in 1963) caused a stir in the scientific community. This suggests the continued acceptance of the claim that tool use is a predominantly human activity, and a general lack of interest in the subject of animal tool-use. Indeed,
while scientists recognized that animals had the ability to use various forms
of tools, such activity was seen as innate or instinctive, and not worthy of
consideration in the same scope as human tool-use.

A second point of note is Darwin's implication that tools are used as com-
pensations for biological deficiencies. This argument has been revisited by
a number of scholars. Thus, e.g., in his discussion of Galapagos finches,
Bowman [1] suggests that a twig would be used to compensate for the lack
of pointedness of the bird's beak. Taking this argument further, Alcock
claimed that animals who used tools tended to invade niches in which they
were poorly adapted, and so used tools as a way of compensating for
their '... lack of biological equipment' [2]. For example, if species of birds
with different beaks were to invade this niche, they might not need to use

tools. This notion can be read in two ways: (i) animals invade a new niche, for which they are not evolutionarily adapted, and then develop tool-using skills in order to survive, or (ii) animals with the propensity to develop tool using skills are able to invade new niches. In either case, tool use is seen as an evolutionary imperative, through which successful species adapt to their environment. While this is an attractively simple argument, it does not stand too close a scrutiny. For instance, if one considers related species that live in similar environments, one does not always find shared tool-using abilities. Thus, gorillas have not been observed to use tools in the wild, even though their habitat is often similar to that of apes who do display tool use. Further, failure to use tools in the wild is not an indicator of an inability to use tools. Gorillas have often been observed to play with tools in captivity or can be taught to use tools with a degree of proficiency. Consequently, the notion that tool use is first and foremost a means of biological adaptation to the animals' environment is dubious. Having said this, the vast majority of instances of animal tool-use are directed towards extraction of food from the environment.

In this chapter, we review tool use by animals other than primates. The aim is to develop an overall picture of how different species make use of tools, and also to relate animal behaviour to the forms of engagement proposed in Chapter 1.

Tool use by insects, crustaceans and fish

As mentioned in Chapter 1, historically there has been an association between tool use and intelligence. In other words, it was assumed that a defining feature of humanity was its ability to use tools, and that 'lesser' species did not use tools as a result of their lower intelligence. On any scale of 'intelligence', one would place insects and fish fairly low. If it can be demonstrated that these species have the ability to use tools, then either we separate intelligence from tool use, or we revise our estimates as to how clever these creatures are.

Table 2.1 presents a summary of some activities in which insects, crabs and fish use external objects to effect changes on the environment. In each case, the creature carries or holds the object, using whatever means

Table 2.1 Instances of tool use by insects, crabs and fish

Species	Object	Function	Activity
*Ant lion	Sand	Throwing	Flicking with head
Myrmicine ants	Leaves	Carrying	Carry in mandibles
Mud wasp	Stone	Hammering	Carry in mandibles
Marine crab	Anenomes	Brandishing	Holds in pinchers
*Archer fish	Water	Throwing	Shoots water at prey

available to it. In three of the cases, the creature uses its mouth (or mandibles) to carry or project the object, and in one case, the creature uses its claws.

I have marked two of these examples with an asterisk (*) because the activity somewhat stretches the notion of a 'tool' proposed in Chapter 1. Whilst both instances, ant lion and archer fish, demonstrate the ability to change the environment, the activities involve using an element of the environment (rather than the manipulation of objects to effect change). Having said this, both instances involve the creature 'projecting' a substance (sand or water) at its prey. The fact that the substance is not produced by the creature, but is taken from the environment and reused represents what might be termed *proto-tool-use*. In other words, the animal is manipulating material from the environment in order to make a change to objects in the environment. Alcock [3] refers to the ant lion activity as an example of preadaptive behaviour, i.e., the insect already possesses the propensity for head tossing, and so the movement of sand could be seen as an incidental consequence of aggressive activity. The archer fish seems to exhibit the water-shooting behaviour as an innate response, and does not seem to benefit from experience [4], i.e., it does not get more accurate with practice or age. Thus, while it is not clear what behaviour is being shown in shooting water, it is probable that this is neither a learned nor adaptive behaviour.

It is interesting to speculate on the origin of the other activities shown in Table 2.1. For the marine crab, the claws are used to manipulate food, so one could imagine that the anemones are grasped in a similar fashion (although movement away from the mouth is obviously a modified behaviour). In a similar manner, dresser crabs will pick up material from their surroundings, using their claws, and drape this material over them to provide camouflage. The myrmicine ant is able to carry prey in its mandibles, and so is adapted to carrying objects. In this respect, using the object to carry food signals a modification of the object to be carried.

A further point to note is that the creatures reviewed so far tend to exhibit very specific behaviours, and for these behaviours to be performed in very specific environments. In other words, it might be the case that for the mud wasp, tamping with a stone forms part of the overall activity of burying eggs in a burrow. What is not apparent is whether these creatures perform these activities in other contexts. In other words, it is difficult to determine whether these activities are learned responses, whether they are innate responses to specific stimuli or whether they are part of specific action sequences.

Having said this, there is clear evidence from the examples reviewed to suggest that tool use is context sensitive, i.e., it tends to indicate interactions between predisposition, environment and adaptive behaviour. One possibility is that the behaviours represent S–R pairings, with some form of environmental reinforcement of specific activities supporting the continued pairing of activity X with a particular objective.

From the review in this chapter, it would appear that the animals typically show specific morphological characteristics that enable specific forms of engagement, e.g., the ability to grasp and manipulate objects in beaks, mouths, mandibles and claws. In other words, the animal is able to accommodate a particular object. The objects are available in the environment, and the animals make no attempt to modify them. Indeed, one might say that the animal combines object and environment into a single entity, upon which it can operate. This implies a form of environmental engagement, in which the animal manipulates the object and environment in parallel, together with morphological engagement, in which the animal's body can act upon objects.

What is less apparent is the sense in which the animals show goals. Do the animals represent goal states or do they have action sequences that are elicited by characteristics of the environment? For example, does the mud wasp have a sequence of actions associated with laying of eggs that involves making a burrow, laying the egg, covering the burrow and tamping with a stone – or does the mud wasp decide to tamp down the earth with a stone? My feeling is that the former is more likely to hold. In other words, whilst the examples reviewed in this chapter demonstrate animals' ability to manipulate objects, it is not clear that such behaviours involve any form of engagement requiring intentionality. However, this assertion presupposes a particular view of intentionality, i.e., intention as pre-planned schema for action. An alternative perspective could view intention as being a component of the action itself, i.e., rather than generating a 'script' based on an understanding of the situation, the animals could simply act and through their actions display an intention. As we shall see in Chapter 5, the comparison between intention as something pre-planned and intention as a component of action is equally valid when applied to skilled human tool-use. Consequently, it is not necessarily the case that the animal tool-uses reviewed are 'unintentional', but possible that they merely exhibit a different form of intentionality. Having said this, the goals are so closely intertwined with the environment and the tool to make it difficult to isolate specific components; in other words, it is not easy to see how the manipulation of tools would be performed if the environment were changed, which, in turn, implies such a close coupling of environment–tool–goal that it is difficult to believe that the animal represents a 'goal' in any meaningful sense.

These examples raise specific questions pertaining to tool use that will be revisited throughout this chapter (and subsequent chapters). The first question relates to the physical ability of the creature of perform a specific action. It would appear that the tasks require specific morphological engagement. However, this does not appear sufficient for tool use as creatures with similar morphologies do not exhibit such actions. The second question relates to the role of the environment (or objects in the environment) in eliciting the action. The third question relates to the question of intentionality.

In their discussion of animal tool-use, Parker and Gibson [5] point out the distinction between tool use that is 'context specific', rather than 'intelligent'. I prefer not to use the word 'intelligent' in these discussions; it seems to me that the word is overly loaded with expectations of higher-order knowledge and information processing. From the discussion of forms of engagement in Chapter 1, the examples discussed demonstrate environmental and morphological engagement. Thus, tool use might represent a particular S–R pairing, in which a given object in the world evokes a given response from the creature, or it might be cognitive, in which the pairing of object and response is mediated by some goal-directed, purposeful activity. It is noteworthy that the creatures considered in this section do not appear to generalize their responses to different contexts. Further, if tool use was an S–R pairing, then one might anticipate certain additional features of S–R learning, i.e., acquisition of the S–R pairing, but there has been few if any reported sightings of these creatures not being able to perform the action and then, through reinforcement, acquiring it. From this, we are initially forced to concede Beck's [6] point, raised in Chapter 1, that tool use merges with myriad other activity. The morphological engagement arises as a consequence of the creatures' ability to manipulate objects, and the environmental engagement arises from the presence of such objects. Finally, while the majority of activities here relate to either acquiring food or defence, the breadth of species makes it difficult to see clear-cut links between tool use and particular characteristics.

Tool use by birds

There has been a wide range of coverage in the popular media concerning the intelligence of birds. In a common format for television programmes, common garden birds, such as tits and sparrows, are set 'tests'. The 'tests' typically involve nuts placed into tubes, with sticks preventing access to the nuts; the birds need to remove the sticks in order to make the nuts drop down the tubes. Certain species of birds (and some squirrels that discover the tests) are able to solve the problems and to retrieve the nuts. A striking illustration comes from David Attenborough's *Life of Birds* television series. Crows waited at a pedestrian crossing (on a university campus in Japan). When the traffic stopped, the crows placed walnuts, fallen from nearby trees, on the road. When the cars moved away, the birds flew up to the trees and waited. The cars drove over the nuts, cracking them open and the crows waited until the traffic stopped before flying down to collect the opened nuts. One explanation of this behaviour hinges on two ideas: (i) the crows already know that they can crack hard objects by dropping them. It might be the case the walnuts simply bounce when dropped, so the original behaviours are thwarted; (ii) the crows observe that cars smash walnuts as they drive over them. They might pair 'moving cars' with 'nut cracking'.

Such behaviour, while not tool use *per se*, covers several aspects of behaviour that are of interest to this book, i.e., the ability to manipulate physical objects, the ability to generate goals and plans to reach these goals, the ability to recognize an appropriate pairing of action to objects, the ability to respond appropriately to environmental opportunities. What these activities imply is that the birds are able to take a relatively simple set of activities, and to apply them in increasingly complex patterns of behaviour. Of course, one might feel that the manipulation of sticks with the beak are very much innate abilities, e.g., consider the activities involved with nest building or feeding. However, the examples presented earlier imply that the action itself is less significant than the manner in which the action is employed and the capability to string sequences of simple actions together in pursuit of a goal.

While it is possible to artificially construct scenarios in which birds (or other animals) are encouraged to exhibit tool-using behaviours, it is more interesting (and I feel valid) to consider such behaviours in their naturally occurring states. Thus, Table 2.2 shows some of the reported natural uses of 'tools' by birds.

Vultures and gulls exhibit similar propensities for using solid objects to crack eggs or shellfish. For gulls, the primary behaviour is to take the shell in the beak, fly up and then drop the shell. Of course, the height is crucial in this endeavour: too high and the shell will shatter and the insides will be lost, too low and the shell will be unaffected. Dropping hard objects onto solid ground appears to be common behaviour amongst several species of gulls, crows and other carrion birds.

The vultures, rather than dropping eggs onto hard surfaces, will use hard objects to impact on ostrich eggs (Figure 2.1). Thus, the vultures will use a stone, held in the beak, to hammer or throw at the egg. It is worth noting that the vultures will also crack pelican eggs by smashing them on the ground. The implication is that, rather than developing a novel response, the vultures are modifying an existing response; in other words, the use of stones to crack eggs could be viewed as the reorganization of the actions used to crack eggs on stones. This explanation could also be applied to the performance of the crows using cars to crack walnuts.

Table 2.2 Instances of tool use by birds

Species	Object	Function	Activity
Egyptian vulture	Stones	Hammering	Hold in beak to hammer or throw
Galapagos finches	Twigs	Probing	Hold in beak
Brown-headed nuthatch	Bark-scales	Prying	Carry in beak
Sea gulls	Rocks	Throwing	Drop eggs onto rocks

Figure 2.1 Egyptian vulture using stones to crack open an ostrich egg.

According to Alcock [7], the use of stones to crack eggs (by vultures) is typically 50 per cent accurate, i.e., the hammer or throw is as likely to miss the egg as to hit it. The activity is relatively slow in that some 4–12 hits will occur over 2–8 minutes. This suggests that the bird is not only impacting the egg but also spending some time checking the results of the action, e.g., hitting the egg, checking to see if the egg has opened and, if necessary, repeating the action. This process is analogous to the test-operate-test-exit (TOTE) model of action proposed by Miller *et al.* [8]. The TOTE model proposes that actions are guided by plans, and the plan draws upon an actor's knowledge of the world. The matching between a 'goal' and the current state requires some feedback mechanism which checks whether the goal has been reached. Thus, as Figure 2.2 shows, the TOTE model assumes that actions run through a simple process of checks and acts.

For the example of crows presented earlier in this chapter, it was clear that if the walnut was not cracked, it would be repositioned. Thus, the crows can also appear to be following a TOTE process. Interestingly, the egg-cracking behaviour of Egyptian vultures can be easily elicited by using 'fake' eggs, which suggests that the behaviour represents some form of S–R pairing. Furthermore, the Egyptian vultures that exhibit this behaviour will often feed with white-backed vultures who have not been observed to exhibit egg-cracking. This supports the earlier assertion that similar species in the same environment need not show tool-using behaviour (in other words, this example reduces the proposal that the environment somehow creates tool using).

Table 2.2 also shows the well-known use of twigs by Galapagos finches and the, perhaps lesser known, use of bark-scales by the nuthatch.

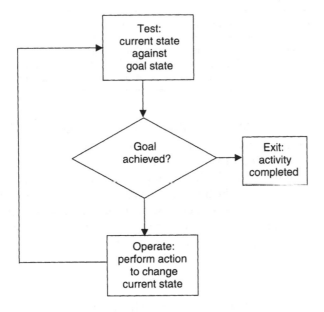

Figure 2.2 TOTE.

The nuthatch breaks of a bark-scale and then uses this as a lever to remove other pieces of bark from a tree to expose insects. The finches have been observed to break off twigs with their beaks, and then to insert the twigs into cavities in trees in order to probe for insects. The twig either impales the insect or acts as a bridge for the insects to walk across. What is particularly interesting is the observation that the finches break off the twigs themselves (as opposed to simply picking up dropped pieces of twig). This implies that the finches are able to 'make' the tools they need, i.e., by breaking twigs to beak size. However, it is not clear whether the finches 're-use' the 'tool', i.e., once the insects have been coaxed from the hole, the finch will drop the twig in order to use its beak for feeding. Also, the widespread observation of the Galapagos Woodpecker finches using twigs to catch insects suggests that this activity is largely innate, i.e., a genus adaptation to the environment rather than a learned or individual adaptation. Recent observations of New Caledonian crows also indicates propensity for tool use amongst birds [9]. The crows have been observed manufacturing and using hooked tools, made by plucking and stripping a barbed twig, and also a 'stepped cut tool' with serrated edges, made from leaves. Both tools are used to extract insects from crevices.

It remains a moot point as to the level of cognition involved in these activities. It is apparent that the birds are often capable of defining a goal

Table 2.3 Instances of tool use by mammals

Species	Object	Function	Activity
Elephants	Branches	Scratching	Hold in trunk
Polar Bears	Rocks	Throwing	Hold in claws and throw
Sea otter	Stones	Hammering	Carry in paws and hammer

and performing actions that will achieve that goal. This implies an ability to separate environmental, morphological and motor engagement in a manner that is not exhibited by the insects, fish and crabs considered earlier. Furthermore, the fashioning of tools, from twigs and leaves, and the use of stones as hammers, implies that the birds have insight into the functions of objects as tools to help them achieve their goals. Indeed, recent accounts of a Caledonian crow in captivity, tell of its ability to form a piece of wire into a hook that it can use to retrieve food from a container.

Tool use by mammals

As noted at the start of this chapter, Darwin pointed out that elephants would use twigs and branches to swat away flies, although Goodall [10] noted that the elephants were more likely to use branches for scratching themselves. In these examples, the branches are tools, in the sense that they are physical objects used in the pursuit of a goal, albeit an ill-defined goal (i.e., reduction of discomfort).

A well-known example of tool use by mammals is the sea otter's use of stones to crack open shells.

The sea otter balances a stone on its chest, and then pounds a shell against the stone. The rate of blows is quite rapid, i.e., about 2 per second. The typical pattern of activity is to strike the shell against the stone a number of times (1–15) and then to test the shell, by biting it, to see if it has cracked open. The process of hammering and checking can be described by the TOTE model in Figure 2.2.

The only other example of mammalian tool-use that I have uncovered is the use of rocks by polar bears. The bears will throw (or drop) rocks onto seals or walruses during attack.

Table 2.3 shows instances of tool use by mammals.

Motor engagement: preadaptive or goal-directed?

Table 2.4 summarizes the activities covered in this chapter in terms of motor engagement. The actions are fairly simple and tend to follow a similar pattern, i.e., grasp objects in beak, claw, etc. and then move object in rapid actions with limited control. The actions imply that the object becomes a weighted extension of the limb holding the object. This supports

Table 2.4 Summary of tool-use behaviours

Function	Object	Number of occurrences
Throwing	Stones, rocks, sand, water	4
Hammering	Stones	3
Scratching	Branches	1
Prying	Twigs	1
Probing	Twigs	1
Carrying	Leaves	1
Brandishing	Anenomes	1

the earlier proposal that, for some animals, tool use represents a form of preadaptive behaviour that builds upon existing patterns of activity.

For sea otters (and crows and finches), a more complex picture emerges. First, the level of motor engagement can be more controlled, particularly in the case of crows and finches removing insects using sticks. Second, the motor engagement is demonstrably performed in pursuit of a goal, with the animal checking the outcome of actions in order to determine that the goal has been achieved. Third, the fact that the objects are selected and worked upon suggests that some 'value' is attached to the object. In other words, there seems to be perceptual engagement at play here; the animal responds to the stone not as something *in situ*, i.e., as part of the environment, but as something independent of environment and much more related to a particular goal. The notion of intentionality is then likely to arise from both cognitive processes of stone selection (and transporting stones between sites) as from the motor forms of engagement involved in manipulating the stones.

Discussion

The focus of this chapter has been on tool use across several species. An interesting point to note is that tool use appears to be the exception rather than the rule amongst animals. The point of this observation is that the ability to use tools does not seem to be merely an adaptation to the environment or an innate predisposition; if it were, then it is not clear why other species in the same environment, or the same species in different environments fail to display such behaviours. With the possible exception of the birds and sea otters, the behaviours appear to be part of action sequences and it is not easy to see that the animals represent either the goal of the action or the nature of the object in a manner to support adaptive and flexible behaviours.

For the most part, the activities reviewed can be considered in terms of morphological engagement between animal and object, leading to motor engagement in which the object is applied using existing actions. Environmental engagement relates to the selection of the object and to simple checks on the outcome of the actions. The sea otters, crows and finches all appear to define goals, and to relate their activity to these goals, but it remains unclear as to the level of cognitive and cultural engagement that is at work here. The issue to consider is how do cognitive forms of engagement manifest themselves; in other words, what are the cognitive dimensions of tool use? In order to address this question, Chapter 3 will consider the use of tools by primates and human infants.

3 Tool use by primates and young children

An anthropomorphous ape, if he could take a dispassionate view of his own case, would admit that ... though he could use stones [for] ... breaking open nuts, yet that the thought of fashioning a stone into a tool was quite beyond his scope ... Nevertheless the difference in mind between man and the higher animals, great as it is, certainly is one of degree and not of kind.

(Charles Darwin, 1871)

Introduction

While one might assume that 'tool use' is a characteristic of primates, observations conducted in the wild tend to suggest that only a limited set of species exhibit such behaviours. Thus, while primates might be considered significant users of tools, it is not necessarily a defining characteristic of many species of primates in the wild. For instance, orangutans might display limited tool use in the wild but are very adept tool users in captivity. Furthermore, even within particular groups, tool use might vary. Thus, one group of chimpanzees might use sticks as weapons while another might not, or one group of chimpanzees might use tools for acquiring food, while others might not. This suggests that tool use, in these examples, is not simply a matter of a species' abilities (as these can vary according to all manner of contextual factors), nor simply a matter of environment (as similar species do not all respond to the environment in the same manner). In other words, we need to extend consideration of forms of engagement from the basic environmental-morphological-motor pattern that characterizes some animals, and include other forms of engagement. In this chapter, it will be proposed that, like some of the birds considered in Chapter 2, primates exhibit perceptual engagement during tool use, but that, unlike birds, there is ample evidence for cultural and cognitive engagement as well.

Tool use by chimpanzees in the wild

Jane Goodall reported observations of chimpanzees using tools in the wild in the 1960s [1]. Since then there have been many observations of such behaviour. Table 3.1 summarizes a few of the studies that have been reported. The letters refer to specific sites covered by specific reports. My aim in producing Table 3.1 is simply to show the breadth of tool-using activities that have been observed amongst chimpanzees. I have divided the behaviours into specific activities. The object is the particular thing used, and the focus is the purpose or goal of the activity. I have also classified these behaviours into specific classes.

From Table 3.1, it is clear that there is a wide spread of activities, performed by chimpanzees in the wild, that can be construed as forms of tool use. However, not all chimpanzees show all of these activities. The most common activities, in Table 3.1, relate to some form of display, e.g., throwing, flailing and clubbing sticks and branches. Other activities seem to depend on the availability of food, e.g., fishing for ants or termites, cracking nuts, etc., and on some level of culture, i.e., tool use is restricted to specific populations and shared within that population. Clearly, the chimpanzees are using objects found in their immediate vicinity to forage for food in the immediate vicinity. As we shall see in Chapter 4, the objects might undergo some form of modification, e.g., leaves may be stripped from twigs, or selection, e.g., stones may be selected on the basis of their weight. This suggests environmental engagement, in the availability of artefacts, morphological engagement in the selection of artefacts, perceptual engagement, in recognizing the appropriateness of objects and selecting and modifying these objects, sophisticated motor engagement in their use of the objects and cognitive engagement in the ability to select an object as a means to achieving a goal. Finally, of the 45 behaviours in Table 3.1, around one-third (15 out of 45) are concerned with finding food, i.e., extracting or foraging. These activities, rather than those related to aggression or display, are proposed to exhibit characteristics that can be associated with the notion of tool use being developed in this book.

Extractive foraging

In this section, studies of extractive foraging by chimpanzees will be reviewed. The aim is to illustrate the range of activities that have been observed in naturally occurring behaviour.

Ant dipping [2] has been observed to occur with a mean duration of almost 20 minutes, with insertions at around 30 second intervals. This suggests both that the chimpanzees were persistent in their actions and that they repeated their actions in order to attain specific goals; in this case, having ants to eat. At one level, this activity might be dismissed as a form of operant conditioning, i.e., the intermittent feedback (ants on the stick) was sufficient to reinforce the activity of dipping the stick into the anthill. However, the regularity of the action, i.e., at 30 second intervals, and the actions involved, suggest that some other explanation is more plausible.

Table 3.1 Observed tool-using behaviours amongst chimpanzees in the wild

Activity	Object	Focus	Class	A	B	C	D	E	F	G	H	I	Total
Fish	Stick	Termites	Forage	✓		✓			✓				3
Dip	Stick	Ants	Forage	✓				✓	✓				3
Dig	Stick	Termites	Forage								✓		1
Fish	Stick	Ants	Forage	✓		✓							2
Probe	Stick	Bees	Forage				✓						1
Pick	Stick	Marrow	Forage				✓						1
Gouge	Stick	Gum	Forage							✓			1
Hammer	Stone	Nuts	Hammer/ Forage	✓	✓		✓						3
Club	Branch		Hammer	✓	✓	✓		✓					4
Flail	Branch		Display	✓	✓	✓		✓				✓	5
Sponge	Leaf	Water	Carry	✓	✓		✓						3
Haul	Branch		Carry		✓								1
Napkin	Leaf	Self	Groom	✓								✓	2
Groom	Leaf	Self	Groom	✓		✓							2
Swat	Stick	Fly	Groom									✓	1
Toothpick	Stick	Self	Groom									✓	1
Scratch	Stick	Self	Groom									✓	1
Leaf-clip	Leaf		?		✓	✓							2
Throw	Stick	Self	Display	✓	✓	✓	✓					✓	5
Tickle	Stick	Self	Play	✓									1
Play	Branch		Play		✓	✓							2
Total				11	8	8	5	3	2	1	1	6	

Key to sites: A: Gombe; B: Bassou; C: Kosije; D: Tai; E: Kanta sui; F: Assinik; G: Kanton Sapo Tuvai; H: Capo Okorobilko; I: Wamba. McGrew, A-H; Ingmanson, 1996 – Wamba.

Viewing film of chimpanzees performing these actions, one cannot help but anthropomorphize their actions and make assumptions relating to their feelings of success, effort, frustration, etc. This is probably an inappropriate response to the activity, but does tend to imply that behaviour demonstrates a high degree of purpose. From a more objective perspective, it is clear that the activities involve sophisticated motor engagement. For instance, the chimpanzees hold the base of the stick in a firm grip using teeth or foot, and then grip the middle of the stick between thumb and index finger in order to manipulate the stick and move it around inside the hole. The stick is manipulated in the hole for a period of time and then extracted quickly, raised to the mouth and licked. This bimanual operation signifies a well-developed form of motor engagement, and the required coordination suggests a learned skill.

A further point to note is that chimpanzees will select sticks, for ant or termite extraction, according to definable principles, i.e., the diameter of the selected stick is proportional to the hole to be dipped – too small and the insects might escape, too broad and it will not fit [3]. One interpretation of this is that the chimpanzees have a 'model' of the type of artefact required for different contexts.

Using the term 'model' implies, I think, a cognitive approach to the problem, i.e., the chimpanzee might be assumed to calculate the required diameter of a stick and then to search for one to meet this requirement. Indeed, some researchers have proposed that it is necessary for chimpanzees to, somehow, mentally represent the relationship between the tool, the action to be performed and the goal. There are two reasons why I find this proposal unconvincing. First, there are no reports of chimpanzees picking up sticks that 'might be useful'. In other words, there does not seem evidence of chimpanzees selecting material for future use. From this, I deduce that the chimpanzees do not need to hold a representation ('model') of 'useful sticks' and do not need to use such a representation when they are not engaged in termitting (or ant dipping). Compare this with the tendency of people, when walking along a beach, say, to pick up stones or driftwood and to assign some 'meaning' to the found object, e.g., this is pretty, this would make a good flagpole for my sandcastle, etc. My proposal is that a cognitive account would view the sticks as possessing 'meaning'. An alternative account would be that a stick 'affords' a good fit into the hole, and that matching of the stick's properties with the hole's constraints could be learned after a few years of trial and error. This is not to deny that the ability to 'see' whether a given stick will fit a given hole is not a skill, but suggests a different set of processes than would a cognitive account. By analogy with human activity, one can imagine 'seeing' whether a table would fit through a doorway; it is not necessarily to cognize the nature of the table or the size of the door, but it is important to perceive the relative dimensions of these objects. This notion of affordance will be explored in more detail in Chapter 5, when discussing the nature of craftwork. This does not mean that ant-dipping need not involve cognitive engagement, e.g., the chimpanzee needs to recognize that using a stick is an appropriate way of achieving a goal. However, given that ant dipping is often performed by a group, there remains the question as to whether this represents a cultural engagement, in that the group possesses knowledge about ant dipping that members acquire. In this respect, one could propose that the chimpanzee is repeating the activities of other group members, rather than engaging in cognitive activity of its own.

An impressive (although less commonly reported) behaviour is the extraction of honey from a bees' nest in a tree [4]. In this behaviour, a chimpanzee was observed to use no fewer than five tools. A stick chisel was used to break into the bees' nest in a tree. The chimpanzee would grip the stick and bang against the nest. Next, a smaller chisel was used to widen the gap. Following this, a pointed stick (held in the teeth and hand) was used to puncture the inside of the nest. Finally, a stick was used to dip into the nest and extract the honey. The level of intentionality involved in this sequence of tasks is very interesting, as is the use of a 'tool set' used to support the various tasks. This example suggests stronger support for cognitive, rather than cultural, engagement in that the chimpanzee has developed and perfected a set of activities through which the main goal is decomposed into subgoals and each subgoal attacked using different tools. This need not mean that the chimpanzee was able to plan the activity to any great extent beforehand, but rather that the activity could be structured in a manner that allows efficient response to changes in the environment. Thus, the activity could have proceeded through trial and error or through importing known activities, such as hammering and probing, to a new situation. As mentioned earlier, one of the notions of cognition in this book is that cognition involves the ability to shape one's world through one's actions, and the chimpanzee in this example would appear to be demonstrating such behaviour.

Finally, one of the best-known examples of chimpanzee tool-use is in the hammering of nuts, using stones or pieces of wood, to remove the kernel. In one study [5] the chimpanzees were observed to first gather together a collection of nuts and then carry them to a raised tree root. A nut is placed in an indentation in the root, and the chimpanzee hammers the hard shell until it splits. The selection of a hammer (a stone or a branch) seems to depend on the hardness of the nut to be cracked, with different species of nuts requiring different types of hammer. The hammer typically weighs between 1 and 5 kg. The chimpanzees will spend long periods of time, up to 2.25 hours a day over the four months when the nuts are in season. Suitable stones (or pieces of wood) can be scarce in the forests, and chimpanzees appear to memorize the location of (up to five) stones, in order to retrieve one that is near a given tree. Stones can also be transported across the site, although it is not clear whether (or why) stones would be transported between sites; if the chimpanzee knows the location of a suitable stone near a specific tree, it is not necessary to move other stones. The observation that chimpanzees can recall the location of stones at specific sites implies that the chimpanzee attaches value to the stones as instruments for cracking nuts. Chimpanzees, like many other animals, appear to be very good at locating buried or seasonal foods and may well have a good memory for location of such foods; the positioning of stones, therefore, could form part of their memory for the activities associated with eating nuts.

In terms of motor engagement, nut cracking is most efficiently demonstrated by adult females. This suggests that the activity involves a set of skills that take some years to acquire and develop. The idea that nut cracking represents skilled activity further hints at a level of cognitive engagement, in order to coordinate complex patterns of activity. As Vygotsky observes, '...intelligent behaviour as expressed in tool use, is first of all a particular way of acting upon the surrounding world...' [6].

It is worth noting that most observations of nut cracking (and other foraging examples of tool use by chimpanzees) focus on female members of a group. It is also worth noting that it is the females who instruct younger members of the group in these practices (see later). An interesting explanation of this observation is that the females need to find more food than males, because the females have high demands for nutrition and food when they are pregnant, nursing young or sharing with young offspring, whereas the males would forage solely for themselves [7]. The fact that females use tools and share this knowledge with their young suggests a far higher level of cultural engagement than observed in other animals. Before considering cultural engagement, I will focus on tool use by other primates in the wild and also the use of tools by primates in captivity.

Tool use by primates in the wild

Many species of primates throw or brandish sticks, i.e., macaques, gorillas, baboons, capuchins, orangutans and chimpanzees. Capuchin monkeys have a high degree of prehensility and thumb opposability, so are able to manipulate artefacts relatively well. However, most of the observed tool-using amongst this species is confined to the use of sticks as weapons, e.g., to club snakes or other attackers, or as a means of pulling food towards them. They have, however, been observed using stones to crack nuts, and pounding fruit against trees, as have baboons (see Figure 3.1). However, compared with the breadth and sophistication of chimpanzee tool-use, other primates tend to show more mundane tool-using behaviours. It is a moot point as to whether the tool use of capuchins and baboons is 'deliberate', in the sense of being performed with anticipated outcomes, or whether it is the result of the destructive foraging practices of these primates. In other words, it might be that the nut cracking observed amongst capuchins differs fundamentally from that observed amongst chimpanzees. What is interesting in these

Figure 3.1 A capuchin monkey using stones to crack open nuts in Brazil, and a chimpanzee using a piece of wood to crack open nuts in the Congo.

Source: *BBC Wildlife 21 (2).*

observations is that these species would appear to share many morphological and environmental attributes with chimpanzees, and yet do not employ objects in anything like the same manner.

Tool use by primates in captivity

Many of the activities that were reviewed could be characterized as relating to subsistence, grooming or aggression – although many instances of the 'aggression' activities could be as much to do with playing as with any real need to cause harm or challenge a rival. The notion of 'play' implies that the animals could engage in tool use even in the absence of direct reward. Thus, it is interesting to consider what happens when subsistence needs are removed, i.e., when these animals are placed in captivity?

Two studies have compared large numbers of captive animals on simple object-manipulation tasks. For example [8], when zoo animals are shown objects that were new to them, capuchins and chimpanzees tend to show the widest range of responses to these objects, particularly in terms of manipulation. Such studies suggest that the primates were more likely, than other species, to play with the objects.

Köhler's studies

One of the most famous publications on primate tool-use is Köhler's descriptions of the behaviours of captive chimpanzees on Tenerife, during the First World War. His work is best known for introducing the concept of 'insight' to theories of problem solving. For Köhler, *Einsicht* could be thought of as 'insight' or 'intelligence' and was related to the grasp of the structure of a situation based on the interconnection of properties. In other words, when solving problems the chimpanzees would arrive at solutions that were '...complete wholes which may, in a certain sense, be absolutely appropriate to the situation' [9]. It is worth noting that the 'solutions' were not always correct; indeed, Köhler speaks of 'good errors', in which the chimpanzees fail to solve the problem but perform an action that has a certain logic to it. For example, a group of chimpanzees were attempting to open a heavy iron door, but a large stone was blocking the door's movement. The chimpanzees then attempted to lift the door over the stone. At one level, the action is absurd and shows limited understanding of the basic mechanics of the door. At another level, the action addresses the problem of how to make the door avoid the stone. What is interesting about these observations is that they tend to reflect some of the notions of cognition presented in this book, i.e., the chimpanzees were able to define a problem and seek ways to act upon the world in order to solve this problem. It is irrelevant whether their solutions were correct or not, but essential to note that they were making associations between current state of affairs and a goal state, and using their knowledge of objects to define appropriate actions.

Köhler's studies typically took the form of presenting a chimpanzee with a reward, usually fruit, that had to be obtained by overcoming an obstacle. Thus, he might place a banana on the outside of the cage housing the chimpanzee such that it is just beyond the reach of the outstretched hand. The chimpanzee would then succeed in obtaining the banana by using a stick.

As noted earlier, the practice of using sticks to reach for food has been observed in several primate species in the wild. What is interesting is Köhler's observation that the chimpanzees were able to join two sticks together, e.g., they might push a thinner bamboo cane into the top of a fatter one in order to produce a longer stick (see Figure 3.2). This adaptation of the object, i.e., production of a new tool, is not something that is generally observed in the wild. In later observations, another researcher showed that the majority of captive chimpanzees in the study (31 out of 48) spontaneously joined two sticks together, even when there was no reward [10]. This suggests that the joining of sticks is as much a form of motor engagement as cognitive. Indeed, whilst Köhler's notion of 'insight' suggested that the chimpanzees joined the sticks as a consequence of their meditation on the problem, it is equally plausible that the joining of the sticks (through manipulative play) suggested a solution. This suggests to me a form of perceptual engagement in which the properties of the object 'afford' achieving the goal.

Of particular significance for the discussion in this book is the question to what extent do Köhler's studies indicate that the chimpanzees

Figure 3.2 Sultan joins two sticks into a single tool.

are using 'models', i.e., representations, of the artefacts they employ? An interesting aspect of Köhler's work is that the chimpanzees could be thwarted in their problem-solving. For instance, if a piece of fruit was placed outside a cage and one stick placed on the floor of the cage facing the fruit and a second stick placed in a different part of the cage, the chimpanzee would not collect both sticks to join them; it was as if the stick near the fruit was part of the 'problem space' and the other stick was not considered. This suggests that one ought to be a little careful about overly applying the notion of 'model' or representation to chimpanzees' tool-using behaviour.

There continues to be a debate as to whether Köhler's chimpanzees solved the problem through a sudden realization of the solution (and Köhler is incredibly anthropomorphic in his discussion of the animals' behaviour, even down to their moods and facial expressions), or whether the solution was arrived at through trial and error. It is worth noting that even Köhler's own writing suggest that the 'insight' was often followed by further periods of trial and error, implying that if the chimpanzees had arrived at a solution, they had not consolidated (or learned) it. Vygotsky noted that ' ...the extremely limited nature of the life of representations is a characteristic trait of the chimpanzees' intellect and ... these animals as a rule already turn to a blind mode of action when the visual situation becomes just a little bit unclear or optically confused' [11]. The implication is that the chimpanzee does not appear to be able to restructure the space of the problem (i.e., to decide to fetch another stick to join to the one near the fruit). Thus, the level of cognitive engagement that chimpanzees apply in tool use is primarily focused on the *immediate* interaction between themselves, their world and the goals they have set.

However, chimpanzees do appear to demonstrate other levels of cognitive engagement, e.g., in their ability to generalize solutions to problems, and to assign 'meaning' to the tools they use. Vygotsky proposes, '...the use of tools presupposes an understanding of the objective properties of things' [12]. For chimpanzees, it is important to be able to develop functional associations between objects and goals. Köhler noted that from his observations '...the stick...acquired a certain functional or instrumental value in relation to the field of action under certain conditions and...this value extended to other objects that resemble the stick, however remotely...' [13]. Thus, chimpanzees attempted to perform 'fetching' activities using twigs that were too weak or even pieces of string. This suggests the ability to form associations of object with goal, albeit in a manner that could lead to problems.

Primate and human infant development

Seymour Papert describes tools as 'objects to think with', not just to use (the notion of 'things to think with' was propounded by Levi-Strauss and will be developed further in Chapter 7). Papert describes playing with gears in his childhood in an attempt to understand the mathematical ratios that governed them, '...playing with gears became a favorite pastime. I loved rotating circular objects against one another in gearlike motions...I became adept at turning wheels in my head and at making chains of cause and effect: "this one turns this way so that must turn that way, so..."' [14].

It is generally assumed that primates attain sensorimotor proficiency faster than human infants. Prehensile activities can occur at different ages for different apes, e.g., squirrel monkeys demonstrate motor skills at 10 weeks, capuchin monkeys are able to demonstrate varied precision grips and some bimanual actions at around 13 weeks, which is younger than chimpanzees and much younger than humans. Furthermore, capuchins are able to form combinatorial actions, such as dip for honey, at around 12 months, while chimpanzees typically perform similar actions, such as termitting, at 3 years.

Primates are less quick to attain preoperational thinking (around 4–8 years for the apes and 2 years for humans). For example, in one study [15] capuchin monkeys are presented with a perspex tube with a peanut placed halfway down a tube. Retrieving the peanut requires the use of a stick. After 2 hours, 3 of the 6 capuchins tested could use the stick. All of the children over 18 months who were tested could solve the problem with ease. Another very common task, is to give a rake (or L-shaped stick) to primates with a bowl of fruit. After some practice, the primate might be able rake the fruit from the bowl. In another study, hamadryas baboons, after some 12 hours of trial and error practice can perform the task proficiently [16]. Young chimpanzees seem to be quite poor at this task, with one study showing only 2 out of 6 chimpanzees (aged around 4 years) able to perform the task at first. However, after a period of 3 days, during which the chimpanzees were allowed to play with the stick, all 6 chimpanzees performed the task in less than 20 seconds [17]. This emphasizes the importance of play, in terms of developing object manipulation skills.

In order to perform even these simple tasks, it is necessary for the animal to be able to perform fine motor activities, to be able to coordinate motor activities into seamless wholes, to be able to perceive elements of the world as belonging to significant groupings and to associate appropriate motor responses to these groupings. This implies an ability to learn associations (rather than to merely exhibit S–R pairings). Thus, 3–5-year-old children allowed to play with objects show more goal-directed responses and more complex actions than those who had no prior experience [18]. Play seems to support two distinct types of activity: object exploration, in terms of manipulation and acquiring information about the properties of objects, and object play, in terms of learning about the activities that can be performed using objects.

Cultural engagement

For Piaget, the things that children played with were seen as 'neutral' objects, i.e., there was no sense that these things carried meaning other than their physical properties. For Vygotsky, the objects existed as cultural signifiers and were interacted with in terms of cultural norms. In other words, children's play with objects is mediated by the people around them, and these people imbue the objects with significance: both in terms of the naming and meaning of the objects, and also in terms of appropriate responses with those objects. For instance, a bowled object on a handle is called a 'spoon', it is used to carry food from the bowl or plate to the mouth, it is gripped in the hand and is rotated during the movement from bowl to mouth. However, when one learns to use a 'tool', one does not merely learn the appropriate morphological and motor engagements (although these are, of course, essential). Rather, one learns the accumulated knowledge and attitudes of a culture that the tools represent. The tool poses a 'correct' way of working, which represents an accepted mode of operating on the world and the objects it contains. Thus, in the example of the spoon, the bowl of desert spoons tend to have similar dimensions; this might reflect bowl relation to size of human mouth, or the minimum amount of material used for manufacturing processes, but might also reflect particular attitudes to how Western cultures view the activity of eating, e.g., there are various rules of etiquette governing how to hold a spoon, how much food to put onto the spoon and how to eat off the spoon. Furthermore, the grip that is appropriate for a desert spoon being used for ice cream is markedly different from the grip used to hold a soupspoon being used for soup.

Much of what Vygotsky had to say about tools related to the transformation of object manipulation into higher-order cognitive functions, which takes place through 'enculturation'. As we saw earlier, children can easily use tools by 18 months of age, e.g., they can use a stick to reach a toy. Chimpanzees can also use tools to manipulate objects. At first, children appear to behave very much like chimpanzees when trying to solve practical problems, e.g., in terms of trial and error or in terms of manipulative abilities, but from the age of around 2 years, children will *talk* while trying to solve a practical problem. Vygotsky proposed that children use speech as an instrument to organize their behaviour. Indeed, he showed that if children were prevented from speaking, e.g., by having them keep a pencil between the teeth, they were less able to solve problems. Thus, practical intelligence begins to be encultured as soon as children acquire language.

This suggests that the way that we talk about tools can have a profound effect on how we use tools. This idea of a language of tool use is explored in Chapter 7. However, even in the absence of verbal ability, there appears to be a set of components of tool use that can be considered in terms of language, e.g., relating to how one ought to hold a tool, the sort of movements that can be made when using the tools, and so on. In his observations of mother–child chimpanzee pairs during nut cracking, Boesch [19] noted that the mother would attempt a variety of pedagogical strategies. These might involve stimulation of the child's interest, e.g., by leaving the hammer stone at the anvil so that the child could play with it and imitate the older chimpanzees, and facilitation of the child's activity, e.g., the mother might provide 'good' hammer stones and nuts that are not difficult to crack. In addition, the mothers often exhibit active teaching, e.g., the mother might position the nut prior to striking or demonstrates the best orientation of hammer. According to Boesch's observations, such interventions occur with a high regularity, i.e., once every 5 minutes. In this study, the 'language' used related to the structuring (syntax) of the task and to the significance (semantics) of the objects used.

For both human infants and chimpanzees, the acquisition of tool-using abilities appears to involve both object exploration and some form of instruction by an experienced (usually adult) user of the tools. However, it is interesting to note that the process of instruction appears to differ between these groups, particularly in terms of how the learner responds to the instructor. 'For human children, the goal or intention of the demonstrator is a central part of what they perceive and, thus, her actual methods of tool use – the details of the way she is attempting to accomplish that goal become salient. For chimpanzees, the tool, the food, and their physical relation are salient; the intentional states of the demonstrator and her precise methods, on the other hand, are either not perceived or seem less relevant.' [20]

Discussion

The broad conclusion of this chapter is that chimpanzees demonstrate all of the forms of engagement proposed at the start of this book. This suggests that their tool-using capabilities are more sophisticated than other animals. The selection and modification of objects, partly through play and exploration, partly through enculturation, and partly through seeing the relationship between tool and goal, illustrates the breadth of cognitive engagement that can be seen in the behaviour of chimpanzees. The use of environmental engagement, of responding to the affordances of objects in the world, is similar to some of the observed behaviours of birds and mammals. However, the chimpanzee shows an ability to generalize skills across domains (particularly when observed in captivity). This implies a level of understanding of the nature of the objects-as-tools. That chimpanzees are observed to make quite fundamental mistakes in their behaviours suggests that their representations are incomplete and not always appropriately generalized. However, the nature of the mistakes often provides telling support for the proposal that some form of representation is possibly being used. What is less obvious is how one can distinguish between those behaviours that are possibly based upon responding to the affordances of objects, and those behaviours that require some type of representation. Finally, the most striking characteristic of chimpanzee tool-use is the level of cultural engagement through which tool-using behaviours are shared and transmitted within a group.

4 The making of tools

The systematic making of tools of varied types required not only for immediate use but for future use, implies a marked capacity for conceptual thought.

(Oakley, 1972)

Introduction

In Chapter 2, tool use was stretched to include almost any object that is used by an animal to perform a task. It was proposed that tool use in animals is typically related to environmental and morphological engagement. This means that tool-using activities can be linked to both environmental opportunity and to a propensity for the animal to behave in a certain manner, either through innate behaviours or through adaptation. In Chapter 3, the use of tools by primates, especially chimpanzees, was shown to exhibit additional forms of engagement such as cognitive and cultural. In this chapter, the focus will be on the modification of objects found in the environment in order to manufacture tools.

In his work on the human hand, Napier proposed that tool use amongst animals involves, '...an act of improvisation in which a naturally occurring object is utilized for an immediate purpose and discarded' [1]. Such objects have been called 'naturefacts'. Thus, the Mud Wasp uses a stone to tamp down mud over an egg and then discards the stone. The object is not carried beyond the immediate vicinity of use. When the animal moves to a new site, a new object is found and used. Furthermore, it is unlikely that the object obtains any special significance, i.e., it is hard to conceive of the wasp finding a twig or stone and thinking 'that would be useful for mud tamping' and carrying it for future use. In all probability the availability of the object in the environment, combined with appropriate cues, will be sufficient to elicit the 'tool-using' response. The fact that most members of the species exhibit such behaviours in a stereotyped manner further implies that the response is innate.

Other forms of tool use require objects to be taken from nature and changed in some fashion in order to produce 'artefacts'. The two primary means by which artefacts are produced are modification and making.

Chimpanzees, other primates and some birds show a different pattern of tool use to other animals. It was proposed, in Chapters 2 and 3, that in addition to environmental and morphological engagement, many of the tool-using activities of these animals illustrates perceptual engagement. Whilst some of the activities, such as nut hammering, make use of available objects, others involve some change to given objects. Napier refers to such action as tool modifying and describes it as '...adapting a naturally occurring object by simple means to improve its performance: once used it may be discarded or retained' [2]. Modification usually involves some form of reduction, e.g., removing leaves from a twig, but could also involve conjunction, e.g., joining two sticks as in Köhler's studies.

For some writers, tool modification necessitates cognition, i.e., in order to shape the stick for ant or termite fishing, the chimpanzee must 'represent' (imagine or model) the required dimensions. As I proposed in Chapter 3, this is not a necessary argument: it is feasible that modification is primarily a perceptual process, arising from the principles of affordance proposed by Gibson [3]. From this perspective, object relationships in the world 'resonate' with specific cortical structures, enabling the viewer to 'see' patterns of object-action. An often used example relates to the trivial action of opening a door: one approaches the doorknob, recognizes that its spherical form supports a particular form of grasp and reaches out with one's hand in an approximation of this grasp. Further, as one knows how doorknobs turn, e.g., rotate away from the doorjamb in order to open the latch, then the grasp is angled to apply maximum torque for this operation. It is an interesting study to stop oneself when reaching for doorknobs (or taps/faucets), just prior to grasping and looking at the posture of one's hand; the posture is seldom appropriate for simply grasping the object, but is appropriate for performing the action [4]. In other words, the simple act of reaching for an object in order to manipulate involves not merely forming the hand into an appropriate grasp, but also posturing the hand to produce an appropriate effect. Indeed, the 'decisions' involved in shaping the hand seem to be made quite early in the movement. The point that I am making here is that it is feasible for the animal to employ a sense of the affordance of objects, such as twigs, and for tool use to involve environmental engagement, i.e., for the chimpanzee to 'see' that a given hole in a termite mound requires a twig of a certain diameter. This is not to deny that such perception is complex, that it is not learned and developed through experience and practice and that it is not prone to error (all of which indicate that a level of skill and proficiency is required).

Even when animals modify found objects, they always use teeth or hands to make fairly rudimentary modifications. In his studies of captive chimpanzees, Köhler noted that the chimpanzees were able to join sticks together, suggesting a slightly more sophisticated process (i.e., constructing a new tool rather than merely removing leaves), and suggested that this was a cognitive process. However, as we saw in Chapter 3, spontaneous joining of sticks is common in captive chimpanzees and may well reflect motor engagement with the objects, as well as play. The point to note is that, to date, no chimpanzees have been observed using objects to manufacture objects; in other words, there is no evidence of any animal using 'secondary tools' in the wild, apart, that is, from humans.

For Napier, tool using and tool modification are distinct from tool making, which he defines in terms of the following process: '...a naturally occurring object is transformed in a set and regular manner into an appropriate form for a definite purpose' [5]. To this extent, tool making is a development of other tool-related activities. Again, for some writers, tool making implies a high degree of cognitive sophistication. One argument is that tool making distinguishes humans from primates because it involves more 'cognition'.

Making stone tools

In his account of tool making by early hominids, Oakley writes that the trimming of sticks for ant dipping or termite fishing by chimpanzees, '...is a far cry from the systematic making of stone tools... which evidently required much premeditation, a high order of skill and an established tradition implying some means of communication' [6]. While I am broadly sympathetic to this quotation, particularly in terms of the emphasis on skill and culture, I am less convinced by the notion of premeditation, which reflects, I feel, a particular stance of the animal–human distinction.

There is a tendency, in much of the literature on tool making to place too great an emphasis on the cognitive aspects of tool making; in other words, researchers have too readily assumed cognitive engagement without thinking through other forms of engagement. Thus, one might read that stone tools require the maker to have a form in mind prior to working the stone. But this is much the same argument that applies to chimpanzees stripping twigs; it is not necessary to hold an image of the resultant artefact prior to construction. Indeed, as we shall see in Chapter 5, skilled craftworkers do not typically formalize a complete object prior to construction, but are much more likely to work towards a form while responding to changes in the material that they are working upon. In other words, there is far more environmental, motor and perceptual engagement in such work than cognitive engagement. Of course, this proposal hinges on one's interpretation of cognition; if one assumes that cognition involves planning and premeditation on an action, then it should be clear that these actions do not require a great deal of cognitive engagement. On the other hand, one needs to make judgements pertaining to the progress of the work, intermittently during the process, and these judgements will guide, coordinate and control the process.

In broad terms, the making of stone tools involved the removal of flakes from a core, by striking the core with other stones or hard material, such as antler or bone. This is, in my experience, an incredibly difficult activity; knowing how to hold the stone, the angle at which to strike it and how much force to use are all significant challenges to the novice flint knapper. Originally, palaeontologists proposed that the core represented the 'tool' and the flakes were a by-product of the manufacturing process [7]; some of the flakes could be used for cutting and others were waste ('debitage'). The notion of the core forming a tool probably led to the assumption that making stone tools involved manipulating the core to form a predetermined shape, and that this shape was held as a model or template by the toolmaker. However, this account is not necessarily true of all forms of stone-tool making. As Toth and Schick point out 'These core forms are not necessarily tools, nor do they necessarily correspond to "mental" templates held by early hominids' [8]. In order to appreciate what might be happening, experimental archaeologists have sought to reconstruct stone tools and, through this, to determine how stone tools could have been made.

The process of making stone tools typically involves some form of flint knapping. Oakley [9] observed twentieth-century flint knapping, in the south of England, related to the manufacture of gunflints. From his account, one can gain much insight into flint knapping. A large block of dry flint is placed on the knapper's knee. A 5 lb, steel-faced hammer is used to tap the block, and the sound the block makes indicates the plane of weakness in the block. A heavy blow from the hammer across the plane of weakness is sufficient to split the block into four pieces. The knapper then takes one of these pieces and, with a lighter hammer, strikes off parallel-sided flakes in order to form a fluted cone. This requires the knapper to hold the piece in one hand and strike with the hammer, rotating the piece in the hand between strikes. Finally, the cone is hammered on an anvil to produce sharp-edged rectangles.

The points to note from this description are (i) the process involves several senses: hearing, vision and touch; (ii) both tool and workpiece are manipulated, with the two hands working together; (iii) the finished product arises from the nature of the material, i.e., one might not be able to predict how the four pieces will split when the block is struck and one might not be able to determine precisely how much material will be removed on each blow. Thus, the knapping of flint becomes very much a process through which several forms of engagement are at play, with the need to continuously modify performance in the light of changes in the material. This means that, rather than working to a 'template' and a 'plan', the flint knapper will interact with material and tools in an ongoing and continuously changing process. This is not to deny that decision-making is performed throughout the task, with the knapper interpreting perceptions and relating these to the overall goal of producing the gunflints. Reflecting this back to the making of early stones, one can propose that ' ...the ability to envisage such geometric relations as symmetry of platform and cross section, and the ability to create a straight or regular edge... ' [10] become key requisites for making stone tools. My feeling is that an appreciation of how stone fractures and the consequences of hitting across a particular line represents knowledge accumulated through practice and that this knowledge informs and guides activity. Furthermore, such knowledge would seem to be more detailed than that required to use stones to crack nuts; both activities can progress in a TOTE manner (see Chapter 1), but the knapping of flint requires a host of considerations that are not relevant to nut-cracking. Thus, even the most rudimentary of flint knapping can appear to have a requirement for cognitive engagement. However, as we shall see later, cognitive engagement is not a necessary component of such activity. Before examining different forms of stone tool, it would be useful to develop the comparison between flint knapping and nut-cracking with a consideration of whether primates can produce stone tools.

Studies of primates working stone

While there are no observations of primates making tools, or working stone, in the wild, there have been attempts to teach captive primates basic stone-tool making. Thus, in one study, an orangutan was taught to fashion a stone flake that could be used to cut string, in order to open a box and get food [11]. In this study, the core stone was fixed in place, and pre-shaped, so the orangutan did not have to make any decisions about where to hit the core or how to grasp it or even how to wield the hammer; the trick was to hit with the hammer stone until a flake dropped off. Having said this, it is noteworthy that the orangutan was able to produce flakes consistently and that it persisted in the activity (rather than refusing to work with the stone). A more important point to note is that this study indicates that it is possible for apes to use tools to make tools (although this is not something that is observed in the wild).

In a series of studies, a captive chimpanzee, Kanzi, was taught the hard-hammer percussion technique for stone-tool making [12]. The flakes would be used to cut string that held the lid of a box containing food. Kanzi, a 10-year-old *Pan Paniscus*, had participated in many studies relating to language learning. After some nine months of teaching, practice and trial and error, Kanzi was able to produce flakes from core stones. The flakes resembled Oldowan pieces. Before progressing, it is probably worth recalling that *any* controlled impact on stone could produce Oldowan-like flakes and cores, so the question is whether Kanzi was producing 'tools' in any meaningful sense (or whether he had learned to remove flakes from stones)? In broad terms, Kanzi did not seem to appreciate that angle of impact influenced the produced flake, but tended to hammer forcefully and randomly. Indeed, his favourite method of producing flakes was to hurl the stone onto the floor. In her account of his activities, Savage-Rumbaugh tells how she laid carpet on the concrete floor to prevent him throwing the rocks; only for Kanzi to lift the carpet in order to reveal the concrete floor and to resume throwing. In a further development, Kanzi learned to place one stone on the floor and then throw another stone at it to produce flakes.

As the researchers pointed out, Kanzi tended to hit the stones with too little force, with blows angled too steeply (i.e., at 90° rather than at the angle of 75–80° required to produce flakes), and tended to produce small (i.e., <4 cm) flakes. Whilst this might indicate poor performance, it is not obvious that the impairment is cognitive so much as morphological, i.e., the long fingers and short thumbs of chimpanzees make it difficult to grasp the hammer stone in a manner to produce a good, clean glancing blow. Consequently, one might anticipate many of the problems that relate to the substandard flakes he produced. Does this mean that Kanzi was unable to engage with the production of the stones in a cognitive manner? The answer to this question is not immediately apparent. Clearly, Kanzi was able to formulate appropriate solutions to the problem of producing flakes, and could demonstrate several strategies. Furthermore, he was able to consistently apply these strategies. Finally, the production of flakes was directed towards the goal of cutting string to gain access to food. On any measure of cognitive performance, these are clearly indicators of good performance. What seems lacking is the concept of shape of the flakes and, more obviously, the notion of working the core to produce a bigger, stronger and sharper flake. Having said this, it is clear that Kanzi was producing tools that were sufficient for the goal of cutting string, but was not working to a goal of making artefacts of specific shape or appearance. This raises the question as to what sort of stone tools were produced by early hominids?

Types of stone tools

There are different approaches to the classification of stone tools. In this section, I will consider two main types: Oldowan and Acheulean. Whilst I have opted for considering specific ages, it is worth noting that the ability to make specific types of tools tended to cross ages and places, i.e., there is no simple, linear progression from very simple to complex tools.

Oldowan tools

Pebble tools have been dated as being made some 2–5 million years ago. These tools originated primarily in Africa, and some of the best-known examples were found in the Olduvai Gorge. The location gives these tools the name Oldowan. Such tools are associated with *Homo Habilis* (although they could also be linked to australopithecines).

The pebbles, usually of quartzite, were shaped into choppers by flaking in two directions, i.e., hammering alternately at one side and then the other.

Figure 4.1 Pebble tools from the Oldowan period.

A large, unshaped end served as the grip for the hand. Figure 4.1 shows examples of these types of stone tools. Notice that the faces must have been worked by flaking to produce both the pattern of marks and also the working edges. It is possible that some of these pebble tools could be produced by simply throwing the stones at other rocks or onto the floor, in a manner similar to that used by Kanzi. However, the main technique for producing these types of tools appears to be hard-hammer percussion. In this technique, the stone to be worked is grasped in one hand and the hammer stone used in the other hand, in a manner similar to that observed by Oakley. The best shape for a hammer stone is a cobble with rounded surface, possibly ovoid. The size of the hammer must be such that it can be gripped between thumb and fingertips, resting on the fleshy surface at the base of the fingers. In order to obtain a reasonable strike, the hammer needs to be angled so that it hits the surface as acutely as possible. The strike is usually on a surface that is already angled, this means that the resulting strike will produce a sliver or flake from the core stone. The core stone will be rotated in the hand, so that blows will be made to alternate sides.

There is some dispute as to whether the goal of hard-hammer percussion was the worked core or the resultant flakes. It has been proposed, from studies in experimental archaeology, that 'The majority of these core forms can be produced simply by the production of sharp-edged flakes from lumps of stone' [13]. From this, one can infer that the 'cores' were the result of working stone to produce flakes. As Wynn and McGrew point out, 'One of the more direct ways to assess the cognitive ability employed in tool-use is through examination of spatial concepts. The spatial concepts required for Oldowan tools are primitive. The maker need not have paid any attention to the overall shape of the tool; instead, his focus appears to have been exclusively on the configuration of the edges' [14]. This calls to mind the attempts of Kanzi to produce stone tools, with the observation that he was not apparently concerned with the production of a core or a 'correct' flake, so much as the production of usable flakes.

One implication of the studies of Oldowan stone tools is that their production could be a matter of environmental, morphological and motor engagement, with the maker hammering the stone until a flake was produced. Having said this, it is clearly a matter of some importance to know where best to strike a blow, and to have some understanding of the possible fracture mechanics of the stones. However, we do not know how much trial and error was involved, or how many wasted or broken cores were thrown away by the toolmakers. The studies involving Kanzi indicate that crude stones can be produced with little if any of this knowledge, but the Oldowan tools hint at some proficiency and a level of perceptual engagement that allows the maker to determine appropriate places for impact. What is less obvious is what the goal of the activity might be: were the makers producing cores or flakes? One way in which to consider this question is to ask what the stones might have been used for, i.e., to consider forensic analysis of the tools.

Forensic analysis

It is possible to analyse the possible uses of stone tools by using microscopy to examine the wear of stone tools, e.g., in terms of scratches and striations on the tools and on bones collected from nearby, or in terms of damage to the edges and polishing on the surface. The main focus of such analysis is often the blade or tip of the tool, which can show very specific forms of wear relating to the hardness of the material that it was used on and the direction in which it was used. Thus, one could identify cutting motions across material that move back and forwards, such as sawing, or one direction, such as whittling, or cutting motions with the material, such as scraping. From such analyses, it is possible to determine whether the tools were used to remove meat from bones, to separate bone from sinew, to chop and cut plant or wood materials, etc. The suggestion is that different types of stone tool would be used for different purposes, and that early hominids develop 'tool kits' containing sets of tools for these varied purposes. In order to produce different sorts of tools, it is necessary to have a more sophisticated method of production.

Acheulean stone tools

Some stone tools were clearly worked to ensure a high degree of symmetry, using bipolar techniques. In a bipolar technique, the core is held against an anvil stone and the hammer used in a similar fashion as in hard-hammer percussion. The result is a smaller core, often more clearly 'shaped' than other tools and a pitted anvil stone. Such tools are commonly associated with the Acheulian period of some 1 million to 500,000 years ago. These tools are normally linked with *Homo Erectus*. As Oakley points out, 'The most characteristic Acheulian hand-axes...are almond-shaped, oval or

Figure 4.2 Acheulean hand axe.

roughly triangular bifaces with relatively straight margins.... Their surfaces
are formed by shallow, skimming flake-scars' [15]. Some skill is needed to
allow the worker to recognize and interpret appropriate faces for striking
and to determine the appropriate angle at which to strike. This is more than
simply motor skill, and requires a level of perceptual engagement. It also
requires the worker to continually stop and take stock of the manufacturing process, in order to determine what action to perform next. This calls
to mind the TOTE model described in Chapter 2. The decisions required in
the production of Acheulean tools seem to not only relate to where to strike
the core in order to produce flakes, but also how best to shape the core into
a balanced, symmetrical object. This requires deliberate and coordinated
working of the stone, and suggests a great deal of cognitive engagement.
Figure 4.2 shows an approximation of an Acheulean axe.

In the Upper Paleolithic period, stone tools were made into narrow,
parallel-sided flakes that were sharpened along one edge. Generally, a blade
is twice as long as it is wide. These tools tend to be associated with
Mousterian cultures (named after Le Moustier in France). Often multiple
blades were produced from the same core, suggesting that the prime focus
of work was the 'flake' rather than the core for these tools. There is some
evidence to suggest that these tools were produced using indirect percussion
methods, i.e., a hammer stone (or piece of wood or anvil) is used to hit a
stone chisel in order to produce more accurate shaping. Following these
developments in the design of the stone tools, came the introduction of
handles and shafts into which the blades or heads could be fitted. The introduction of the handle seems to me to be a significant turning point in the
making of tools; not only were the blades and heads being produced for
specific tasks, but they were also being modified to fit housings that, in
turn, were made to support particular ways of using the tools. The implication is that the handle became not simply a way of separating the hand
from the tool's head or blade, but also a means of representing the manner

in which the tool was manipulated by providing the opportunity to use movements with a flexible rather than rigid wrist.

Cultural engagement

It is not the place of this chapter to settle any dispute as to what the intention of early hominids was in producing stone tools. It seems apparent that different techniques result in different finishes, both of the core and of the flakes. The variation in technique might have been due to the type of rocks being worked, or to the finished product, or to cultural factors, such as the 'correct' way to work stones in a particular region. As Oakley suggests, 'The primitive hunter made an implement in a particular fashion largely because as a child he watched his father at work or because he copied the work of a hunter in a neighbouring tribe' [16].

In order to produce stone tools, it was necessary to recognize the value of certain materials. Certainly there is ample evidence to show that early hominids transported material to make stone tools between sites and would work the stones in different places. This suggests a degree of transportation that is not observed among primates. Furthermore, the fact that stone tools from different sites are dispersed over wide areas suggested both that these tools were being carried for *potential* use (as opposed to being taken to a specific place for a specific purpose) and, possibly, were involved in trade. From these latter points comes the recognition that the tools themselves possessed value, in the sense that they were useful and necessary. Finally, the observation, from micro-wear analysis, that different tools served different purposes suggest a differentiation in tool manufacture and use that emphasizes a concept of allocation of function, i.e., this tool is suited to this task, and can be fashioned in such a way as to optimize performance of this task.

Tool-making, therefore, represents as much a cultural activity as a physical one. In order to recognize the value of tools, it was necessary for shared goals to be accepted and worked toward. Tool-making would appear to be discretionary, i.e., there is no evolutionary imperative at work here because these peoples would not necessarily die in the absence of tools as they could acquire and eat foods using hands and teeth. What seems to be happening is the recognition that tools are extending the range of functions open to them, perhaps in terms of recognizing the range of foodstuffs available to them.

In terms of division of labour, it might be the case (as implied in Jim Crace's novel *The Gift of Stones*) that not all of the tribe would be involved in tool manufacture, but that certain skilled individuals would be recognized and given preferential treatment. In this respect, tool production would be performed by a subsection of society rather than by everyone. This further distinguishes human involvement with tools from that of other animals; for the animals, most members of a species appear equally proficient at using or fashioning tools. This is primarily true, with the exception of some forms of tool use by primates, where 'skill' is apparent. Not all chimpanzees exhibit tool-using behaviour, but there seems to be variation across groups. Thus, like other activities (such as washing potatoes in sea water before eating them), there would seem to be a strong cultural component to primate tool use. Furthermore, even within a group, not all of the members are able to use the tools with an equal level of proficiency, suggesting some skill that needs to be developed and practiced.

Discussion

The stone tools of early hominids reflect highly consistent production. This might be a matter of the tools reflecting their methods of production. The hands of early hominids were more similar to apes, in that the shorter thumb made anything other than a power grip difficult. The stone tool would be held in the hand, with a flat surface against the palm and applied in a hammering or cutting motion. Alternatively, the tools might reflect agreed notion of what stone tools ought to look like, i.e., culturally defined and accepted objects that would be used in the performance of specific tasks.

A further point to note is that, as tools attain value and as they become associated with skilled makers, so the tool becomes a cultural product. Archaeologists unearthing stone tools can make judgements as to how the users of these tools probably lived. The tools become reflections of ways of living, showing an insight into cultural conventions, e.g., in terms of how the tools are worked, and how the tools are designed to be used. This implies that the tools were reflecting not only how materials could be worked (or how the tribe believed that materials should be worked) but also how the makers of the tools assumed that people would use these tools (or how the tribe believed that work would be performed). Later tools can be associated with specific regions and still later we can distinguish the marks and styles of specific craftspeople.

The production of stone tools reflects levels of cognitive engagement that are not seen in animals. Cognition is not so much pre-planned and schema-driven as opportunistic, allowing for a flexible response to changes in the properties of the objects being worked. Furthermore, the overall goal is not the immediate consequence of using the tool, so much as the production of a tool that can be used for future activities, and that can be reused.

5 Working with tools

The most typical and familiar example of the workmanship of risk is writing with a pen, and of the workmanship of certainty, modern printing.

(Pye, 1968)

Introduction

In this chapter, we turn our attention to the ways in which skilled practitioners use tools. I have taken the epigraph from David Pye, a master craftsman who has written eloquently on the nature of workmanship and craftwork. He draws a distinction between two types of workmanship: that of risk and that of certainty. One way in which one can distinguish the 'workmanship of risk' and the 'workmanship of certainty' is to consider the difference between those artefacts that are made by machine, e.g., through mass production, and those that are made by hand. However, this is a somewhat romantic view of the production processes and does little to help in understanding the use of tools. Many aspects of mass production require some level of manual skill, and a great deal of work done in 'hand-making' things requires the use of machines, jigs and fixtures that aim to both simplify the work and also to ensure greater consistency of work (two characteristics of mass production). By way of illustration, imagine the construction of a piece of furniture, such as a bookcase. Many readers may have bought self-assembly furniture, of the sort that comes in flat packs and is assembled at home. In this case, we use a small set of tools, e.g., screwdrivers, allen keys, hammers, to fix the various pieces together to form a pre-defined unit. The pieces have been machined to specific sizes, holes have been drilled at specific points, the fixtures have been designed to fit in specific orientations and our task is to follow the instructions and join the pieces appropriately. This is analogous to Pye's notion of the workmanship of certainty: the entire assembly process has been designed to minimize variation. Of course, this is not to rule out the possibility of problems arising – how many of us have misread the instructions for self-assembly furniture and had to undo some of the work in order to progress in the correct direction?

In contrast, the workmanship of risk represents almost the opposite approach to working, i.e., '...the quality of the result is not predetermined, but depends on the judgment, dexterity and care which the maker exercises as he works' [1]. The craft worker exercises a whole host of skills that enables the manipulation of tools, the working of materials and the design of artefacts. Indeed, the '...ability to use tools is the traditional criterion of the craftsman's skill' [2]. In his study of industrial skills, Seymour draws several distinctions between the experienced and inexperienced worker. 'First, the experienced worker usually employs "smoother" and more consistent movements...Secondly, the experienced worker operates more rhythmically, indicating that a higher degree of temporal organization has been achieved. Thirdly, the experienced worker makes better use of the sensory data...Fourthly, the experienced worker reacts in an integrated way to groups of sensory signals, and makes organized grouped responses to them' [3].

In this chapter, I argue that it is not simply that the craft worker has 'superior' manipulative skills than other people, nor that they are just 'better' at using their hands than other people, but rather that the craft worker has a different view of the world and the artefacts it contains. In other words, 'People who use tools...build an increasingly rich implicit understanding of the world in which they use the tools and of the tools themselves. The understanding, both of the world and of the tool, continually changes as a result of their interaction' [4]. Thus, when using scissors to cut shapes from material, the dressmaker might be less concerned with the movement and feel of the scissors than with the straightness of the cut or with following the pattern. Even the most mundane of everyday tools can be subject to such effects, e.g., as mentioned previously, when you use cutlery your attention is less on the handling of knife and fork and more on the cutting and moving of food.

From the perspective of 'workmanship of risk', the decision of which tool to use, how to wield that tool and how to coordinate activity become significant components in the success (or failure) of a course of action. A central question for this book is the extent to which these decisions are made prior to performing an activity, or whether they are an integral part of performing that activity. In other words, is cognition an exercise in planning and preparation or is it the moment-by-moment coordination of activity?

Tacit knowledge

In Chapter 1, it was proposed that tools can be 'forgotten' in use, i.e., the user of a tool concentrates on the task at hand and not the tool itself. It is common for craftspeople to speak of feeling the tool being part of themselves; in Chapter 1, a tool was viewed as a physical extension of its user, but the notion of a tool being part of oneself seems to extend this notion much further. In other words, there is (for the skilled practitioner) an intimate relationship between wielding the tool and performing a task. Thus, an experienced carpenter sawing a piece of wood would focus on the cutting of the wood, whereas a novice user of the saw would focus on moving the saw up and down. To take another example, how many of us have been so focused on writing an essay that we have 'lost' the sense of holding a pen or hitting the keys on the keyboard and only been aware of the words that we have written? This suggests that changing emphasis across the forms of engagement differentiates the expert from the novice; in other words, skill need not arise merely from doing an action more quickly or with fewer errors, but from the ability to focus on appropriate forms of engagement (and to have strategies and techniques for by-passing less important forms, or for assimilating these forms into coherent units of action).

Craftwork necessarily involves the worker interacting with objects in their environment, through tools and processes. At one level, this simply reflects the fact that craftwork involves the manipulation of objects to create new objects, e.g., the blacksmith works hot iron into brackets, the carpenter works wood into chairs. Tools are used to effect changes in materials, by modification or addition. Further, craftwork involves the monitoring of subtle changes in objects in the environment. Thus, the baker needs to 'know' the sound that a loaf makes when you tap it.

Consequently, the skilled user of tools needs to attend to both tool manipulation activities and to the changing environment. This raises significant, and I believe unresolved, questions as to how people divide their attention between such sources of information. Some writers speak of 'tacit' or 'implicit' knowledge when discussing tool use, but this often simply recognizes the fact that some knowledge is difficult to put into words. Skilled practitioners can be said to have a variety of ways of representing tool use. This takes up the point developed in Chapter 7, that the 'language' of tool use can take several forms, not all of them linguistic and not all of them focusing on the specific interactions between tool and user. Skilled practitioners employ all manner of 'jargon' to describe the goals and sequences of actions inherent in their work. This means that just because the carpenter says 'a mortise and tenon joint', rather than referring to specific grips and movements of the saw, hammer and chisel, does not mean that a complete set of knowledge is not being expressed, i.e., the phrase 'mortise and tenon joint' probably contains within it a whole host of assumptions covering the types of tools to use, the appropriate grips and manipulations of these tools, the sequences in which the tools are used, the outcome of the tool-using activity and the appearance of the finished joint. This seems to me to suggest that rather than being 'tacit', the knowledge of skilled practitioners has become very densely packed and that unpacking this knowledge represents a challenge, both to them and to the various human sciences that take an interest in how people work.

Forms of engagement

In Chapter 1, six forms of engagement were proposed to cover tool use: environmental, morphological, motor, perceptual, cognitive and cultural. In this chapter, these forms of engagement will be illustrated through consideration of craftwork in different domains. The domains were selected simply for illustration, although the selection also reflects a relative dearth of scientific research into the area of the skilled (or indeed other) uses of tools. The aim is not to provide exhaustive descriptions of these different domains, but to use examples of work in each to develop a general perspective on the nature of skilled work.

Environmental engagement

In an earlier chapter, mention was made of Gibson's notion of affordance. In broad terms, Gibson proposed that rather than an organism and the environment existing as two separate entities, with one acting upon the other, these entities were somehow conjoined into a single dynamic unit. For example, a specific area of the visual cortex responds to lines of a certain angle such that the presence of lines at this angle leads to firing of this part of the cortex. From this perspective, things in the world do not exist as collections of physical properties (as one might assume from common sense or from 'traditional' notions of perception) so much as collections of features that 'afford' (or support) some specific response. Environmental engagement involves the user of a tool responding to the affordances of objects. This would provide, I feel, a valid and coherent account of many aspects of animal tool-use covered in earlier chapters, but also has implications for skilled work. A point, that will be developed in Chapter 11, is that 'affordance' for tools probably reflects more than perception–action linking; rather, one can speak of the affordance of the form of a tool, e.g., a handle is designed to afford a particular grasp, the affordance of the function of an object, e.g., a shoe or a stone affords 'hammering', the affordance of the operation of a tool, e.g., a hammer affords a particular sequence of movements.

It is clear, from observing and speaking to skilled craftspeople, that much of the work relies on them 'just knowing' what to do or when to do it. Ergonomists speak of process control operators having a 'process feel', by which they mean that operators have internalized the dynamics of the process to such an extent that they are able to anticipate the occurrence of events and are able to respond very quickly to changes. For the operators, this might reflect well-developed patterns of action that have arisen from practice, training and years of experience. For craftspeople, this might also reflect a heightened ability to react to the affordance of things. For example, imagine carving a piece of wood using hammer and chisel. For the novice, the shaping of the wood becomes very much a matter of removing wood in order to form a shape, with the concentration being on moving towards an end product through a series of substages. For the expert, the activity becomes somewhat different: not only do skilled carvers speak of the sculpture 'revealing' itself to them, but also they attempt to exploit features of the wood, for example the line of the grain or the presence of knots, in ways that novices simply cannot. For example, in Byzantine ivory carving, 'Manifestations of the grain are irregular in elephant ivory... When they occur, ... they are obviously delighted in' [5]. While this reflects differences in the ability to manipulate tools, with the experts having assimilated basic tool manipulation to such an extent that it becomes 'automatic', it also reflects differences in response to the environment, or the piece of wood or ivory.

From this discussion, environmental engagement reflects the ability of the skilled craftworker to manipulate objects in the world in a manner that allows form to reveal itself. This claim reflects an aspect of Pye's notion of the 'workmanship of risk'; in this case, the risk lies not only in the manner in which the tool could slip, but also (and more importantly) in the moment-by-moment decision-making that allows the skilled practitioner to modify movements and tool selection to best exploit the current state of the material being worked, and to be able to predict the consequences of particular actions. The fact that such decision-making and prediction can occur instantaneously suggests a level of engagement between worker and object that seems to reflect some of the ideas associated with affordance and direct perception.

Morphological engagement

Tools are designed to support particular types of grip. One might expect that the 'common-sense' grip that a tool supports is the one to use. However, consider Figure 5.1 which illustrates the correct grip for a pair of medical needle holders.

The shape of the finger-guards on the needle holders might call to mind scissors, and the untrained user might slide the thumb into one guard and the forefinger in the other. Whilst such a grip would support the cutting motion used for scissors (by supporting movement of the thumb), it tends also to force the wrist into pronation and thus reduces flexibility about the wrist. The appropriate grip, as shown in Figure 5.1, is to use the thumb and

Figure 5.1 Holding needle holders.

the third finger. Such a grip effectively 'locks' the fingers, so that all motion is about the wrist – which is exactly what is required for suturing, i.e., using the needle holders to move a curved needle and draw thread into stitches.

A second example of changes in morphological engagement relates to the use of a screwdriver. Typically, the handle supports a power grip but when one is first fitting a screw, one tends to modify the grip to a precision grip, i.e., using thumb and fingers, in order to guide the screw into the hole and ensure that it is appropriately positioned.

Thus, morphological engagement is not simply a matter of the hand coming to grasp the handle of a tool; rather it reflects the interaction between tool handle and the posture required to perform the task (or part of a task). It is proposed that such interactions become a significant component of the skilled use of tools.

The skilled craft worker is able to wield tools in ways that make it seem as if the tools has become a part of the hand. As Veblen notes, ' ... the work-man adroitly coerces the materials into shapes and relations that will answer his purpose, and in which also nothing (typically) takes place beyond the manual reach of the workman as extended by the tools which his hands make use of' [6]. From this point of view, the morphology of the tool user is partly a matter of the fit of the tool to the hand, and partly a matter of the reach and shape of the combined hand and tool.

In his classic account of industrial skills, Doug Seymour considers tasks involved in a meat processing plant. He was struck by the apparent complexity of the work, but was greatly assisted by one of the foremen who told him that there were only six ways of using knives for the work. Table 5.1 gives an indication of the subsequent analysis that Seymour was able to conduct for these tasks.

Table 5.1 demonstrates that with the knife in her hand, the butchery assistant is able to produce hand–knife motions that follow stereotypical patterns. However, it is also apparent that some flexibility needs to be incorporated into these movements, e.g., gristle is not distributed evenly across all pieces of meat and so some modification and correction needs to be made when this is present. In other words, a skilled user of a tool is likely to employ a set of actions that are similar across situations, and to be able to modify these actions in the light of changing circumstances. A point arising from this, which is discussed further in Chapter 6, is that designing a tool to support a specific morphological engagement, e.g., by angling the handle to support a particular wrist posture, might only be appropriate if the movements are to be highly regular and repetitive.

In addition to relating to the grip that one uses to hold a tool, there is a further consequence of morphological engagement, i.e., the aesthetic or affective response that one has to the tool. To this end, we might consider a consequence of morphological engagement in terms of emotion. Many readers will respond to the proposal that we have 'favourite' tools, e.g., the carving knife that slices easily through meat and sits comfortably in our

Table 5.1 Excerpts from instruction schedule for meat processing (from Seymour, 1966)

Movement #	Goal	Left hand	Right hand	Attention points
I (a)	Remove fat from meat or gristle	Grip joint 1, 2, 3, 4, T at side. Fingers arched and gripping joint at end and behind knife.	Grasp knife between base of 1, 2, 3, 4 and first knuckle intersection with an angle of approx. 80° between 4 and 1, T on back of blade just beyond guard.	Wrist of the R/H moves through 90° clockwise. Slight flick of the knife... Knife must be grasped firmly. L/H might be used for lifting the fat clear of the knife blade.
III	Cutting meat into strips	Grasp furthest edge of meat from body 1, 2, 3, 4 T approx. 3" from point of knife insertion (to the left). Raise slightly and pull upwards and backwards.	Nick the seam at end furthest from body. Only use tip of the knife at an angle of approx. 45° to joint. Grasp knife as per I (a) but when the hand is closed place the thumb parallel to the blade with the tip of the thumb at the knife end of the guard.	Draw knife through, don't saw. Keep wrist up at an angle. Don't put the 1st finger down the back of the blade.

hand, the egg whisk that feels perfectly balanced when we whisk egg whites, or the pen that glides across paper. In the absence of these tools, we might feel that we are just 'making do' with available equipment. That we have favourite tools reflects an emotional response in which the design of the tool so perfectly matches our hands, actions and intentions that we can take pleasure in their use. While this might appear to be a peculiarly human response, there are many accounts of primate interaction with found objects that seem to share a similar root, i.e., the pleasure of simply holding and using something, which many commentators ascribe to 'play'.

In his account of writing, Daniel Chandler [7] provides reports of famous authors who have favourite writing instrument, often relating to the 'feel' of the pen or pencil in their hand or the movement of the pen across the paper or the appearance of the written script on the page. This seems to signify, at one level, pleasure in the physical act of writing, but also reflects the intertwining of the writers with the instrument. Thus, Sartre writes, 'I do not apprehend my hand in the act of writing but only the pen which is writing; this means that I use the pen in order to form letters but not my hand in order to hold the pen ... [M]y hand has vanished; it is lost in the complex system of instrumentality in order that this system might exist' [8]. This quotation echoes the notion that practitioners 'become' the tool they are using, and that one 'forgets' the relationship between hand and tool and sees the tool as an extension of oneself. For writers, the notion of the pen as extension of the self has additional overtones, in that the type of writing instrument might reflect either a different personality or writing style or stage in the writing process. As McCullogh points out 'If you feel satisfaction in using a well-practiced tool, you probably do so on several levels. Tool usage simultaneously involves direct sensation, provides a channel for creative will, and affirms a commitment to practice' [9].

Motor engagement

Motor engagement can vary according to the type of tool being used. For example, consider writing on paper using a fountain pen and ballpoint pen. These different writing instruments might lead to different emotional response, e.g., the fountain pen might feel smoother and the writing style might differ from that of the ballpoint pen. Kao [10] demonstrated that the mean writing pressure with the fountain pen would be about 100 g, while for the ballpoint pen it would be almost 180 g. Thus, the ballpoint is pressed harder against the paper. Conversely, the mean writing time was observed to be around 5 s for the ballpoint pen and some 6 s for the fountain pen. This might reflect the fact that users of the ballpoint pen were less careful in their formation of letters that those using the fountain pen.

In a study of stone-bead manufacture in Khambhat, India, researchers attached accelerometers to hammers that were being used by the workers [11]. Two groups of bead makers were studied: group I had some 30+ years of experience following a 12-year apprenticeship; group II had some 20+ years of experience following a 3-year apprenticeship. Both groups consisted of 6 people. Each worker was asked to produce 80 stone and glass beads during the study. The workers were used to working with stone, but glass was a novel material for them. In the studies, group I produced superior, i.e., more spherical and symmetrical beads, than group II. Furthermore, while both groups were able to adapt their technique, e.g., measured through accelerometry, for the familiar stone, only group I was able to adapt appropriately to the glass. Figure 5.2 shows that both groups modify acceleration when producing large or small beads from stone, but only group I shows similar modification with the less familiar glass.

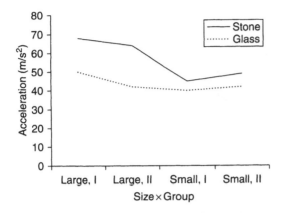

Figure 5.2 Comparison of group I and group II for different materials.

Perceptual engagement

We have already considered some aspects of perception in the section focusing on environmental engagement. In that section, the focus was on affordance and the idea of direct perception. The notion of direct perception is often contrasted with the forms of perception that require interpretation. This is, I feel, a rather tortuous distinction and it is equally likely for the same act of perception to exhibit features of direct and indirect perception. For instance, returning to the example of the baker tapping the bottom of a loaf of bread – the perception of the sound that the bread makes could be indirect, if the sound is not what is expected and requires further interpretation, but could equally be 'direct', if the sound is what was expected and the bread was cooked. Whether the perception really was 'direct' might well depend on the years of experience that the baker possessed, and perhaps for the newly qualified or inexperienced baker, perceiving the sound of the bread would remain a matter of interpretation.

Keller and Keller [12] provide a first-hand account of the skills involved in blacksmithing, by way of a participant-observation study of artist-blacksmiths. The primary means of working in this domain is the transformation of hot iron by working on it with tools. In order to work with metal, a smith needs a knowledge of the temperature of the material being worked. Figure 5.3 summarizes the relationship between temperature, colour, effect on metal and type of working.

Knowing that the metal is at the right temperature clearly requires a well-developed ability to perceive differences in the metal's colour. Whether this perception relates to interpretation and pairing a colour with a temperature with a state of the metal, or whether it relates to simply 'knowing' that a given state of the metal will afford a given type of working, would relate to

	°C	
dazzling white	1500	
white	1400	
light yellow		
yellow		
orange		Forging
light cherry	1000	colours
cherry red		
dark cherry	800	
blood red	700	
dark red	600	
blue heat	500	
light blue		
dark blue	300	
brown		
dark straw	200	Tempering
light yellow	100	colours
	0	

Figure 5.3 Colour of heated metals for forging and tempering (adapted from Keller and Keller, 1996).

the experience of the smith. I once heard an anecdote about steel workers in Holland who were adept at determining whether a length of steel could be pressed by looking at its colour, and could adjust the rate at which the process was run accordingly. The steel works introduced an automated rolling press, in which the metal entered a covered area for working. A consequence of this was that the metal workers were unable to see the colour of the steel, and so lost the opportunity to make decisions on the process. A temperature probe, fitted into the housing, was used to provide a digital display of the temperature, but the workers were not used to responding to temperature as a 'value' so much as a change in the steel's colour. As a compromise, the company introduced a visual display unit that converted the temperature of the steel into a coloured bar that more or less corresponded to the colour of the steel. In other words, the 'technology' was modified to support the worker's perceptual skills by sensing the temperature, transferring this to a computer, converting this to a defined colour and displaying this on a computer screen (when the workers could gauge the temperature simply by glancing at the steel).

It is proposed that perceptual engagement is intertwined with many aspects of environmental engagement. However, when the perceptions require elaboration or interpretation, then the person might approach the task in a different manner. It might be the case that the time taken to perform this elaboration or interpretation would not be appreciable or measurable in any meaningful sense, but might feel like a brief pause between steps in an activity or might lead to the awareness of working with the perception (as opposed to the feel of simply performing the task). The manner in which a course of action is revised will depend on cues from the work. Thus, in other activities, jewellers spoke of metal being heated until it reached a specific colour ('cherry red' rather than 'just red') or that when hammering a flat piece of metal into a shape, the ring of the hammer changed from a 'full' to a 'hollow' sound, or how the scrape of a highly abrasive stone, on a polishing wheel, changed with the angle of the piece altered. Each of these cues suggest a form of perceptual engagement in which visual, auditory and olfactory (e.g., tasting the smell given off by metals under different actions), are used to supplement the motor engagement with tool and workpiece.

Cognitive engagement

Cognitive psychologists often distinguish between knowledge that can be represented as facts, i.e., declarative knowledge, and knowledge that can be related to actions, i.e., procedural knowledge. Keller and Keller speak of two forms of procedural knowledge: technique, as in knowing how to manipulate a tool in order to perform a particular operation, e.g., knowing that a hammer can be used in a variety of ways, using different faces of the hammer or different angles of blow, to produce different effects; and recipe, as in knowing how to combine operations and techniques to achieve an overall goal. The notion of recipe can be related to the concept of script and schema, used in cognitive psychology.

In one of our studies of jewellery making, an experienced jeweller (with some 40 years experience) was observed and interviewed over the course of several weeks. One of the activities that we asked him to perform was a set of 'verbal protocols' (i.e., commentaries upon his actions) whilst making models. The extract presented in Table 5.2 gives a flavour of one of tasks that was observed. The jeweller produced this commentary whilst filing the setting for a stone on a ring. The commentary has been broken down into units of meaning. I have not indicated timings for the statements, but it is probably worth noting that this complete exchange took some five minutes, with lengthy periods of silence in which the jeweller filed the ring. Filing typically took for the form of around 7 to 12 strokes, a pause during which time the jeweller blew filings from the ring and looked at the resulting shape, and then a repetition of the filing. During this time, the ring was held in a vice and the direction of filing was in a straight line across the length of the ring, i.e., the file was moved linearly back and forth perpendicular to the jeweller.

The description begins with a general statement of activity and of the problem at hand, i.e., whether to produce a shank without a split or curl. The statement of the problem and consideration of options and constraints reflect the previous experience of the jeweller. The 'plan' develops according to a 'visualization' of the finished product (or, at least, some concept of the form of the finished product) and defining an appropriate course of action

Table 5.2 Verbal protocol of jewellery task

Verbal protocol propositions	Topic
You are thinking about what you could be doing all the time	Development and revision of 'plan'
And also this shank is more complicated	Defining constraints
It could have been a simple shank and just come up with what we call a spear point shank you joined on there	Considering options
But this shank has got a wraparound and divided or split around the collet, so that's got to be split and wrapped around the collet	Defining constraints and selecting option
You've got to have the right angle and finger size	Defining constraints
You know, all the time you are checking to see that the angles are symmetric as it should be	Checking and revision of plan
Once again it is still checking and checking, bearing in mind that the setter has got to let it out a few thou more	Defining constraints and considering consequences of actions

that will progress from the current state to the finished product. For example, when interviewing another jeweller, who was cutting material for necklaces from a sheet of metal, it was apparent that the main constraint was not efficient use of metal (which could be resold at face value) but ensuring that one cut finished with sufficient space to allow a subsequent cut to be made.

Of particular note from Table 5.2 is the fact that the 'plan' is continuously developed. Figure 5.4 shows a sketch that the jeweller drew of the proposed work. Note the relative lack of detail and the somewhat limited information it contains. This sketch suggests not a blueprint for a design so much as a suggestion for a course of action.

What is apparent from this initial discussion is that the 'plan' that guides the work of the jewellers we have spoken to is not a prepared, precise list of actions (as one might assume from the word plan) so much as the rewriting of a course of action based on current options and constraints. This is very much akin to the notion of 'situated action' that Suchman [13] constrasted with traditional notions of planning. In other words, the intentionality for the action is not something that is predetermined but something that arises from working (see Chapter 1).

Cultural engagement

A central thesis in this book is that the purpose of tools is to allow their users to act upon their environment, in order to achieve specific goals. The ideas of Vygotsky, alluded to earlier, have recently found expression in Activity Theory. One of the foundations of this theory is that of tool

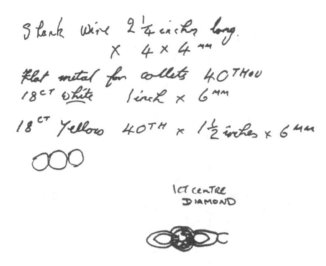

Figure 5.4 Jeweller's sketch of ring.

mediated activity. In other words, tools mediate the relationship between tool users and their environment. Engeström [14] proposed a schematic through which to consider Activity Theory. A version of this is shown in the Figure 5.5.

In Figure 5.5, the 'Mediating Artefact' signifies the tool being employed. This is related to two components: the subject, or user of the tool, and the object, or the purpose for which the tool is being used. This represents a simple form of some of the observations made previously that tool use was purposeful behaviour. Activity Theory then proposes that both subject and object are influenced by cultural factors. Thus, there are rules by which tools are used. These rules might be simple expressions of 'best practice', e.g., someone skilled at using a saw might tell a novice to rub the blade with candle wax to prevent it sticking.

The rules and conventions to which the practitioners in a domain adhere, represent shared beliefs, values and expectations concerning how that work is to be done. One could consider the layout of the workers' workplace as an example of the culture of the work. Thus, an operating theatre tends to be laid out in a particular manner, not only to provide access to the patient and space for the equipment but also to allow communication between members of the surgical team and to provide easy access to relevant equipment and instruments. Indeed, since the 1950s, surgery has been supported by specific kits of instruments that are put together for particular types of operation. In their analysis, Keller and Keller observe that the layout of a smith's workshop reflects the tools that have been acquired to do the work,

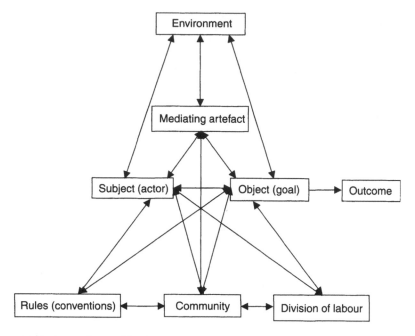

Figure 5.5 A simple model of an activity system.

and the layout of these tools in relation to specific tasks. In other words, the collection of tools and their layout represents an aspect of knowledge of the tasks at hand. The basic layout relates to the need to heat metal, hammer (or otherwise work) that metal and quench it, with equipment related to these actions being grouped together and equipment related to peripheral actions, such as drilling, being moved to other locations. Thus, the layout of the workplace implies a particular way of working. What these examples share is the notion of segmenting the work into constituent components. For blacksmithing, the crucial aspect is working on the metal at the right temperature, and working before the metal cools. This means that tools have to be ready to hand, that the course of action needs to be established and that the smith needs to coordinate activities as rapidly and coherently as possible. A similar example of workplace layout can be seen in Figure 5.6. Tools are grouped according to function, size and process. At the centre of the bench is a block against which rings and other objects are held for filing, and at the right of the jeweller a gas jet is used for softening the metals being worked.

Rules convey the experience of previous users of the tools. Indeed, it is plausible to assume that the relationship between rules and subject/object influences the design of the tools themselves; in other words, tools represent the distillation of centuries of experience in the design of specific artefacts.

Figure 5.6 A jeweller's workspace.

The rules are held by a community. The community might be a troop of chimpanzees sharing rules relating to the cleaning of potatoes in the sea or a group of blacksmiths sharing knowledge about how to beat metal. The community both holds the rules and forms ways of expressing and passing on these rules. The community might also seek to limit the passing on of rules. Thus, guilds of the Middle Ages sought to keep some aspects of their working practices secret. In the Industrial Revolution, one of the motivating forces for developing new ways of managing work was on the division of labour through which employers sought to withhold knowledge of some jobs from employees, and thus weaken the power of the guilds.

The notion of division of labour calls to mind two issues. The first, political issue, relates to the design of work. For example, in the 1770s a manufacturer of pearl buttons in Birmingham (John Taylor) was reportedly breaking work into very simple, discrete operations, i.e., receive oyster, open oyster, remove pearl, clean pearl, polish pearl, prepare pin, prepare head, mount pearl on head, mount head on pin, etc. Similar job designs were reported by Adam Smith and Charles Babbage. The explicit justification for this was that work became more efficient. The implicit justification, of course, was that division of labour removed the skill from individual workers and gave it to the managers of work, which meant that unskilled (i.e., very cheap, expendable) labour could be used to perform very simple jobs.

One of the most famous exponents of division of labour was Frederick W. Taylor, whose notion of Scientific Management became a byword for exploitation of workers in the early 1900s. While this is not the place to explore Taylor's theories, it is worth noting that in parallel with simplifying jobs, Taylor expended a great deal of energy into the redesign of tools and workplaces. Thus, for example, in the famous 'pig-iron shovelling' study, Taylor not only redesigned the manner in which 'pig-iron' (unprocessed iron ore) was moved from the yard to the furnace, but also modified the design of the shovel's head in order to improve shovelling load (and reduce spillage). Thus, division of labour can also have implications for the design of tools; indeed, one could suggest that the intention to divide labour, modifies the manner in which a community functions and the rules which govern the use of tools. With the advent of automation, the simple jobs could be mechanized, and the labour further reduced to serving machines [15].

A second, technological, issue also has bearing on this discussion. Ergonomists speak of allocation of function, by which they mean the sharing of activities (functions) between people and technology. The primary justification for using technology relates to efficiency, timing, repeatability, etc., and a secondary justification relates to industrial relations, i.e., robots do not strike. Allocation of function often rests on the mistaken assumption that one can draw up HABA-MABA lists; in other words, one can explicitly state which functions 'Human Are Better At' and which functions 'Machines Are Better At'. However, if one considers which functions are genuinely better performed by humans or machines, one typically arrives at a graph like the one shown in Figure 5.6.

Notice, that there is a large spread of activities that can be performed equally well by humans or machines. In this spread, the design to automate is less a matter of efficiency and more a matter of other concerns. To take a mundane example, assume that you are a postgraduate student about to do some teaching. You are offered an hourly amount (let us be generous and offer you £21.25 per hour) and have an electronic calculator on your desk. First you are offered 9.5 hours of teaching and you use the calculator to multiply the hourly amount by 9.5. Now you are offered an additional 0.5 hour. Do you repeat the calculation using the calculator, or do you simply move the decimal point one place in the hourly amount? The point of this example is that people might allocate functions to tools when they (the people) are capable of performing the task more quickly. The decision to allocate might be economic or political (in the case of industrial work), but is equally apparent in everyday life. To take a second example, a sheet of paper is to be divided into two pieces. Do you simply rip the paper with yours hands, or use a pair of scissors, or use a sharp knife and a straight-edge? Your choice of technique depends partly on what is to hand, but also on other factors, such as what will you use the pieces of paper for? In other words, allocation of function is dependent on the goal of the activity.

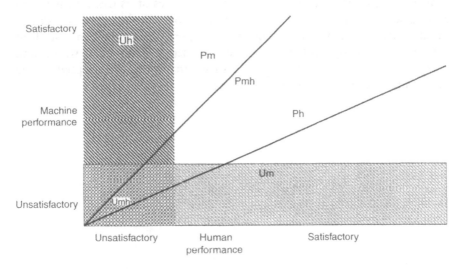

Figure 5.7 Human/machine allocation of function.

How do forms of engagement relate to the notions of Activity Theory shown in Figure 5.5? Clearly many of the forms of engagement relate to the interaction between subject (user), mediating artefact (tool), environment, object (goal). In terms of single links, environmental engagement links the environment to the subject and morphological engagement links the mediating artefact to the subject. However, some of the links are double, e.g., motor and perceptual engagement link the subject to both environment and mediating artefact, and some of the links are triple, e.g., cognitive engagement links subject to object, environment and mediating artefact, and cultural engagement links the rules, community and division of labour to subject, mediating artefact and object. Thus, the forms of engagement attempt to operationalize some of the links that are shown in Figure 5.5.

Discussion

The main thrust of this chapter has been to explore some of the characteristics of skilled users of tools. In particular, I have taken some of the (admittedly limited) reports of skilled tool use in the scientific literature and mapped this onto the notion of forms of engagement. The concept of skill, in this context, becomes ' ... a combination of the sensitivity to visual and tactile input from the ongoing work, muscular control that allows the efficient production of a desired end, and sophistication of procedures implemented to carry out the work' [16]. Much of the work involves the use of tools in a 'free-hand' manner, i.e., in terms of what Pye refers to as 'risk', in that the control of the tool is left to the capabilities of its users (rather than having some form of constraint to limit action). Rather than the skilled practitioner simply demonstrating superior performance in terms of all of the forms of engagement, what we tend to see is a flexibility within and across each form, such that performance can be rapidly adapted to the requirements of the task, tool or material.

6   The design of tools

To be of real value... tools must be designed with full cognizance of the operational system (and its limitations and constraints) into which the device will fit.

(Sanders and McCormick, 1992)

Introduction

To speak of 'designing' tools might be a misnomer; most of the tools that we use on a day-to-day basis have undergone a lengthy evolution to arrive at their current state. The notion of evolution implies competing designs and the survival of some of these designs. If we think of a simple tool, such as the shovel, there are literally hundreds of variations on the basic design, e.g., with varying lengths of handle or size of blade, with each design being associated with specific types of work. One driver for the evolution of tools is the manner in which people continually adapt the design of tools to fit the work that they are doing. Some of these adaptations need not involve specific changes to the tools themselves, but changes in the manner in which the tools are held or manipulated, e.g., a tool with a long handle might be gripped towards the head for some jobs. These changes in use might then find their way into physical modifications of the tools, e.g., the handle might be shortened for this type of work. This suggests a continual process of using tools, changing their manner of use, and then modifying the tools to accommodate the new manner of use.

A second driver in the evolution of tools is the fit between user and tool. As mentioned in Chapter 5, some tool designs seem to better fit the way that a given user performs a task (and many people have 'favourite' tools, that just *feel right*). When people either make their tools (as is often the case with blacksmiths), when they have tools tailored specifically for them (as might be the case for professional musicians), or when they can select from a range on offer (to make a personalized tool kit of favoured tools), then the tools can be said to be tailored for specific individuals to suit both their stature and style of use.

While these points might ring true, they overlook the significance of the mass production of tools. One could assume that, prior to mass production, tools would have been made on a local scale, reflecting the immediate demands of the local environment (e.g., in terms of available materials or in terms of the types of material on which the tools would be used), and the local traditions of working (although this does rather overlook the importance of trade and the movement of tools between different locations). The point is that a spade that was made for digging peat in Ireland might look significantly different from a shovel used to move coal in a mine in Germany. The tools would be customized as tradition, materials and task demands required. Obviously a core tenet of mass production is the notion of standardized products. To this end, the production of shovels would be geared towards producing items that are of a common size and shape, made from materials that satisfy a number of demands, e.g., ease of manufacture, cost, appearance, durability, etc. This results in an approach to tool production that tends to the assumption that 'one size fits all people and all tasks'. Thus, if one enters a hardware store and examines the range of screwdrivers on offer, one will find variations in head design (to accommodate different screws), price, colour, handle design, etc., but all with fairly common features and dimensions. The tools on sale reflect compromises between the tasks that the tool is intended to perform, the manufacturing process employed, the appearance of the tool (which might need to reflect a corporate profile), etc. A significant question, therefore, is whether the standard dimensions of mass-produced tools adequately satisfy the variation in stature of the prospective users?

Despite the compromises between the various factors that influence production, the appearance of a tool reflects received wisdom concerning how the tool is to be used, i.e., a theory of use. It was noted in Chapter 4, that the production of stone tools seemed to reflect a particular view as to how to hold the tool and how it ought to be manipulated, possibly reflecting different theories of use. It was apparent, from considering stone tools, that dividing tasks between different tools led to collections of specialized tools. As we saw in Chapters 2 and 3, chimpanzees and some birds have also been observed to use tool kits consisting of two or more objects, with each object having a specific purpose. The notion of purpose-specific tools implies a theory of use that is based on a fit between tool and task. For example, when one buys self-assembly furniture, one often has purpose-made fixtures that can be used with just a screwdriver and Allen Key. If the fit between tool and task is not sufficient, then one might either seek an alternative tool (e.g., resort to pliers to turn a particularly stubborn bolt) or seek to modify the available tools (e.g., use the pliers to provide additional torque on the Allen Key) or seek to modify the task (e.g., dispense with the idea of turning the bolt and use a hammer instead). In order to consider the physical fit between tool and user, we need to take a short detour into the realm of anthropometry.

Anthropometry of the human hand

It is important to appreciate the physical characteristics of the people who are intended to use the tools, and also to be in a position to consider how poor tool design can lead to problems. Second, it is important to understand how tools are actually employed, in order to propose designs that can support task activity. This section begins with a brief discussion of the anthropometry of the human hand.

The main types of tools considered in this book have been objects that have been designed to be held and manipulated using the hand. Having some notion of the dimensions of the hand provides a useful reference point for the design and sizing of tools. Figure 6.1 shows a picture of an adult male hand (the author's), with some dimensions marked on it. The dimensions are shown in Table 6.1.

Discussion of tool use might also consider how the musculature of the hand and arm support controlled manipulation of tools, and how problems in tool use can lead to injuries to this musculature.

Properties of tools

A characteristic of a well-designed tool is that it feels comfortable and balanced when held. This is partly a matter of ensuring that the handle can counterbalance the weight and length of the blade, partly a matter of the grasp point and partly a matter of material from which the handle is made. When the tool is being used, then additional considerations relate to the

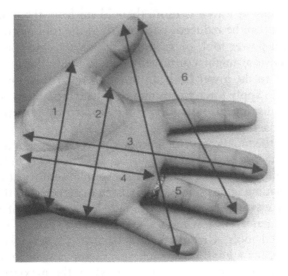

Figure 6.1 The author's hand.

Table 6.1 Anthropometric dimensions for lengths in Figure 6.1 and lengths of digits

Dimension	Male 5th %ile	Male 50th %ile	Male 95th %ile	Female 5th %ile	Female 50th %ile	Female 95th %ile
1	97	105	114	84	92	99
2	78	87	95	69	76	83
3	173	189	205	159	174	189
4	98	107	116	89	97	105
5	178	206	234	165	190	215
6	122	142	162	109	127	145
Thumb	44	51	58	40	47	53
Index	64	72	79	60	67	74
Middle	76	83	90	69	77	84
Ring	65	72	80	59	66	73
Little	48	55	63	43	50	57

mechanical output from the tool, e.g., in terms of the force and torque applied through hand tools or through the power and vibration from powered tools.

In general terms, a tool consists of a handle that one holds and a head that performs a task. The tool is (usually) provided with motive power and controlled by the user. The handle, therefore, is the channel through which power is passed to the tool, the means by which control is exerted over the movement of the tool and the conduit through which feedback from the tool is passed back to the user.

When the centre of gravity of a tool is more than 25 cm from the body, then the ability to lift tools can be reduced, particularly if the tool in question is quite heavy, e.g., a power tool, or is long, e.g., a saw. One consequence of this is that the tool might continually slip towards the lower body, resulting in injuries to the lower limb. In order to resist this downward pull, the user needs to apply force to counter the effect, and the strain of holding a tool with a distal centre of gravity can lead to overexertion injuries to shoulders and back (see Chapter 8).

The centre of gravity of the tool is dependent upon the overall weight and distribution of this weight. In a survey of axes in Latvia, Drillis [1] noted that the average weight of the axe-head was 1.4 kg. From laboratory studies investigating muscle loading, it has subsequently been demonstrated that kinetic energy (from an axe swing) rises proportionally to the weight of an axe-head, until the weight passes 3 kg, when the angular velocity drops significantly [2]. For powered tools, Greenberg and Chaffin [3] propose a maximum weight of 11 kg for a hand tool, such as a chain saw, but note that, for most users, a weight of around 4.5 kg represents the maximum load for manipulation and handling.

Size and shape of handles

In order to consider the dynamic anthropometry of the hand, imagine bending the thumb and index finger to meet around a cone. If a handle was cylindrical, then the hand could slip down the shaft or the hand might be forced to adopt an unnatural grip to maintain contact with the handle. This suggests that handles ought to approximate a truncated cone. Having said this, such a shape assumes specific types of grip, i.e., one in which the handle is held against the palm with the fingers wrapped around it. Furthermore, the manner in which one grips the handle can be used to transmit force through a tool in several ways, i.e., impact forces (as with using hammers), linear forces (as with using saws), rotational forces (as with using screwdrivers): in each case, the requirement for transmitting force will have a bearing on the interaction between user and handle.

The maximum diameter of the enclosed circle ranges from 43 mm for 5 percentile adult female to 59 mm for 95 percentile adult male [4]. Thus, a handle width of around 40 mm is optimal for a power grip (see discussion of 'grip types' later), and a distance of 40 mm will be sufficient to allow optimal grip strength for squeezing, e.g., when using pliers. Furthermore, the maximum width of the adult human thumb ranges from 15 mm for 5 percentile adult female to 24 mm for 95 percentile adult male; given a standard deviation of 2 mm, a circle in excess of 26 mm would be sufficient to allow finger or thumb to be inserted into a handle, e.g., for scissors grip (in practice, the dimension is assumed to be 30 mm). As a general rule, the thicker the handle the lesser the load placed upon the muscles in the hand (although, of course, one must bear in mind the maximal grip dimensions outlined earlier).

Materials for handles

The most common material for making tool handles remains wood. There are several reasons for wood as a primary choice for handles. Obviously wood can be worked easily and can be shaped to fit different heads. More importantly, wood is relatively elastic, which means that it can absorb some of the pressure and shock associated with impact forces so as not to transmit these directly to the hand. Furthermore, wood has a high thermal conductivity, which means that the transfer of heat from hand to tool makes the handle feel warm and comfortable to use. Wood also has a fairly high coefficient of friction, which means that the handle is less prone to slippage than other materials, even when the hands are wet, and that high torque can be exerted with minimal discomfort to the user. On the negative side, wood can be expensive to buy and to work (at least in mass quantities), not always environmentally friendly and prone to damage. The attachment between tool head and wooden handles is a point of weakness, which means that the handles can become detached from the heads.

If other materials are used for the handles, e.g., plastic or metal, then it is important to ensure that they mimic some of the positive attributes of wooden handles. Thus, the metal handle needs to provide sufficient absorption of impact forces, e.g., through the use of rubber coverings, and needs to provide similar coefficient of friction, e.g., through some other synthetic sheathing.

Many tool handles incorporate some form of textural variation, e.g., in the form of rough surfaces, ridges and flutings. This can provide an increase in coefficient of friction in order to minimize slippage. Some handles have indentations for the fingers. However, such indentations (together with ridges and flutings) are very poor in ergonomic design and possess a number of risks to users, particularly related to pressure injuries. For instance, if the indentations are larger than 3 mm, then the morphology of the hand means that many users might suffer compression damage from pressing the fingers across (and into) the gaps. It remains a moot point as to how useful the provision of texture on tool handles will be, at least in terms of actually using the tool. Figure 6.2 shows the results from a study comparing handle diameter with torque exerted by male participants. The data suggest that the prime factor in defining the torque that someone can exert is the diameter of the handle rather than the finish.

Handles in use

The dimensions of handles can also be considered in terms of simple models of how the handle might be used. Thus, one can consider rotation, thrust and impact forces.

Rotational (turning) force (Ta) can be defined by the following equation:

$$Ta = G\mu D$$

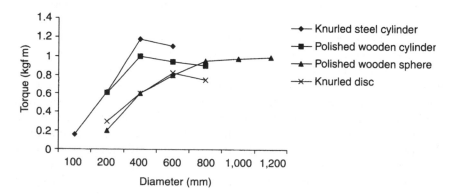

Figure 6.2 Relationship of torque exerted to handle diameter (adapted from Pheasant, 1989).

where G is the grip force exerted on the handle, D is the handle's diameter and μ is the coefficient of friction between hand and handle.

Pheasant proposes that the well-known assumption that the torque generated by a screwdriver is proportional (or at least related to) the length of the shaft is not substantiated [5].

For thrusting actions, such as found in sawing, the relationship between handle diameter and action is defined by the equation:

$$Fa = G\mu$$

where G is the grip force and μ is the coefficient of friction between hand and handle.

Whilst there is no maximum length for tool handles, Drillis [6] proposed that the length of a handle should be at least equal to one 'thumb-ell', i.e., the distance from the edge of the hand to the tip of the thumb. From Figure 6.1, this dimension is similar to the maximal functional spread, i.e., dimension 6. In terms of minimum length, it is worth noting that handles of 19 mm are unsuitable for any tool.

For impact forces, the 'efficiency' of the hammer, i.e., the optimal translation of moving the hammer to the force at the head, is given by the equation:

$$\text{Efficiency } (\eta) = 1 - 1.5\,S/L$$

where L is the length of the hammer and S is the distance from the centre of mass to the centre of action (see Figure 6.3).

Optimal performance arises when S tends to 0. In this case, maximum translation of force would occur when the hammerhead is held in the hand, i.e., as in stone hammers. However, placing the hand at the centre of action can lead to damage to the musculature of the hand arising from the impact, and can also reduce the amount of force that will be applied through the head. Thus, a hand-axe has high efficiency, in that it translates almost all of the hand movement into kinetic energy, but relatively low force, whereas a metal rod would be some 25 per cent efficient but generate a higher

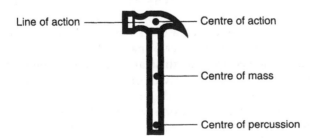

Figure 6.3 Dynamic properties of a tool.

force [7]. The hand is usually positioned towards the centre of percussion, where some of the impact has been absorbed by the handle. Thus, even an action as trivial as holding a hammer would appear to be a trade-off between optimizing impact and minimizing damage. Drillis suggests optimal values of η for different tools, e.g., axes: 0.8–0.95; hammers: 0.7–0.9; hoes: 0.3–0.65.

Hammering is seldom about applying maximum force but about applying sufficient force in a controlled manner. Hence, the handle on the hammer not only absorbs some of the impact, but also allows the impact force to increase. The impact force is proportional to the energy imparted to the head, which, in turn, relates to the mass of the head, the angular velocity of the swing, the distance covered during the swing and the magnitude of the force applied by the user. Generally, the heavier the head, the longer the handle, e.g., a claw hammer might have a handle length of 30 cm and a head mass of 0.8 kg, while a sledgehammer has a handle length of 60 cm and a head mass of 3.3 kg. The longer handle provides a larger arc through which to move the head, so that the swing can be performed using the entire upper body and arms. In this instance, the aim is to produce maximum propulsive force to the head. For smaller hammers, the aim is to produce a controlled impact, and the distance between head and hand provides a separation of the impact and movement to allow better hand–eye coordination.

Using tools: posture, balance and activity

A common approach to considering the ergonomics of tools is to consider the interaction between person and tool from a systems perspective. The system could consist of the user, the tool, the workplace and the design of jobs. The implication from such a perspective is that the design of tools is not simply a matter of ensuring good coupling between hand and handle. Rather it is important to consider how the tool is to be used, both in terms of task requirements and also in terms of the external factors that might influence the task, such as work rate and workplace layout.

Some work activities require the user to adopt postures that are potentially harmful. For instance, using screwdrivers when the arms are raised above the head. In such activities, there is significant strain on the muscles around the shoulder. This results in a decrease in strength, e.g., as measured through maximum voluntary push, and performance does not appear to improve if the person is allowed to hold their arms at rest at their sides between actions. It would seem that the advice of 'resting' between overhead screwdriving actions does not prevent the possible build-up of muscle fatigue. Consequently, a more sensible solution would be to eliminate overhead screwdriving by modifying the workplace, e.g., by using ladders to raise the worker or by modifying a car assembly line to angle the car to provide better access.

A key ergonomic criterion in using tools is to keep a straight wrist, i.e., to minimize ulnar deviation. It is possible to define various 'natural' resting positions of the hand when it is holding tools. For the screwdriver, for instance, the wrist will be abducted some 15°, such that the blade of the tool is some 78° from the horizontal (see Figure 6.4).

Figure 6.5 shows the results of a classic study that demonstrates how modification of a tool's handle, in this case a pair of pliers, led to significant reductions in reports of carpal tunnel syndrome, tenosynovitis and epicondylitis ('tennis elbow').

Recently, ergonomists have turned their attention to the design of hammers. It is well known that when using a hammer, the wrist moves from extremes of adduction to abduction, and that the impact forces can be very high. It has been demonstrated that a handle with a 10° bend is preferred by users (at least for activities lasting for around 5 minutes), that ulnar deviation is reduced and that grip forces can be maintained after periods of hammering [8] for a handle with a 20° bend. These studies indicate that

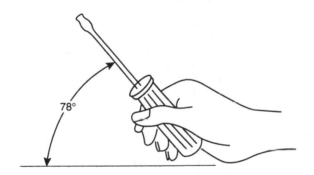

Figure 6.4 Holding a screwdriver (from Fraser, 1980).

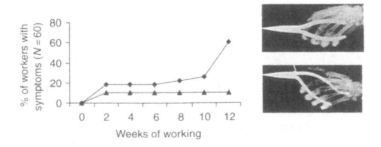

Figure 6.5 Comparison of reports of symptoms for two designs of pliers (adapted from Tichaeur, 1976).

a bent hammer handle not only reduces damage to the user, but also has no detrimental effect on accuracy or power.

Types of grip

A commonly used description of hand grip was originally proposed by Napier [9], and consists of a simple dichotomy between power and precision grips. While such a distinction can distinguish between classes of grip, it does not provide sufficient detail to explore all types of tools. Kroemer [10] proposed a classification of grips based on the configuration of the human hand. Table 6.2 shows the grip types proposed by Kroemer, together with a short description of suggested application. Notice that while the

Table 6.2 Types of grip

Contact	Type of grip	Description	Application
Finger	Finger	Single finger placed on surface; finger either rested or pushed in	Push buttons or touchscreens
Palm	Palmar	Palm placed on surface	Using sandpaper
Finger–palm	Hook	Palm against surface, and fingers hooked around object	Pulling a lever
Thumb–fingertip	Tip	Object held between thumb and (any) finger	Using a sewing needle
Thumb–finger-palm	Pinch	Object resting against palm, and grasped between thumb and fingers	Positioning screwdriver head onto a screw
Thumb–forefinger	Lateral	Object held between thumb and forefinger	Using tweezers
Thumb–two fingers (outside)	Pen	Object rested on thumb and pressed by two fingers	Writing with a pen
Thumb–two fingers (inside)	Scissor	Fingers and thumb placed inside handles	Cutting paper with scissors
Thumb–fingertip	Disk	Thumb and fingers curled around outside of object	Holding sanding block
Finger–palm	Collett	Object rested on palm and enclosed by fingers	Holding a ball
Hand	Power	Object rested across palm and enclosed by fingers	Holding a hammer or a saw

'power' grip of Napier is indicated, 'precision' grips can assume several different types.

The amount of force that can be exerted on a handle varies with the grip [11]. Thus, with a tranversal power grip, an adult female exerts around 400 N of force, but with multifinger or lateral pinch (key) grips this reduces to 100 N, and for a pinch grip this reduces further to 40 N. The obvious parameters influencing force would be the amount of contact between hand and handle (with the pinch grip only using forefinger and thumb) and the muscle groups that can be called upon to support the grips. Thus, the transversal power grip makes use of the larger and stronger forearm muscles to generate the forces, whereas the pinch grip is using primarily the muscles in the hand. A consequence of this variation in muscle group is the precision to which actions can be performed. Thus, the use of forearm muscle groups tend to require a degree of tension about the wrist and forearm, whereas the use of finger muscle groups would leave the wrist loose and supple, which leads to greater precision and control of movement. On a related point, the muscles in the forearm are better able to grip tightly than to push outwards, which means that while these muscles can be used to press together the two handles of a pair of bolt-cutters, it would be advisable to provide a spring to release the handles rather than to rely on the hand to pull the handles open.

Basic principles of tool design

The ergonomics literature provides many examples of guidelines for the design of tools. In the following section, some of this work will be summarized. The interested reader can consult [12–15].

One of the more contentious proposals is that the task should, where possible, use special purpose tools. This relates to the discussion at the start of this chapter, when I spoke about tools being tailored to the stature and style of the people who would use them. Typically, one can find a high demand for tailored tools in domains where work activity is highly skilled, or where such tools can results in efficiency gains.

General points

- Minimize the need for excessive grip force;
- Minimize static loading and pinch points;
- Maximize neutral postures;
- Minimize repetitive finger action;
- Minimize device vibration and impact forces;
- Power with motors rather than muscles.

Handle design

- Design handle to maintain neutral wrist posture, e.g., by having a bend in the handle;
- Use appropriate hand diameter for the proposed task;
- Design handle to assist grip, e.g., through use of textured covers;
- Design handles to support the appropriate type of grip for the task and to encourage use of the appropriate muscle group;
- Design tools to be used in either hand.

7   The semantics of tools

A tool is inscribed in your imagination not only as an activity, but also
as a symbol.

(McCullough, 1996)

Introduction

Semiotics is the study of the ways in which meanings are attached to
objects. Generally, a 'sign' is said to consist of a 'signifier', i.e., an object,
image, word, etc. and a 'signified', i.e., a particular meaning that can be
read from the signifier. The signifier and signified together form the sign.
Thus, a signifier that looks like this 📖 on a mobile telephone might signify 'address book' but on a map of a town might signify 'library', and in
shopping mall might signify 'bookshop'. There are three obvious points to
note from this example. First, if one does not know the signified, then the
sign cannot function, i.e., it is not possible to interpret what the object
means. Thus, if 📖 is placed in a novel location, say on the face of a digital watch, then it might not be easy to associate this particular signifier with
a relevant signified, and you might need to seek additional information in
order to develop an appropriate interpretation. This leads to the second
point that associating signifier with signified can often involve some degree
of interpretation. Of course, as we saw in Chapter 1, there are many examples of everyday technology that have stereotypical interpretations that are
learned and shared within a culture. One example used in Chapter 1 was of
the volume control of a hi-fi; people simply know what this control does
and how to use it. A third point to note is that the meaning of the sign will
vary as context changes. This means that people might misinterpret signs by
associating one meaning of a signifier to the wrong context. From this perspective, the most interesting signs are ones in which there are multiple signified meanings, and there is much work on the deconstruction of images
and objects to explore the variety of meanings that are conveyed. For the
purposes of this book, it is interesting to treat tools as signs and to consider
the simple question what do tools signify?

As far as tools are concerned, it is easy to overlook the fact that the majority of tools are used within environments that contain other tools, e.g., the office, the kitchen, the workshop, etc. The meaning of the tools can be considered both in terms of their unique appearance, in terms of their contrast to other tools in the environment and in association with the goals of the person in that environment. For instance, a knife next to a fruit bowl could have a different 'meaning' to a knife on a desk, i.e., in the first instance, the knife is 'for' peeling or paring fruit and in the second instance, the knife is 'for' opening letters. This is not to say that each context only yields a single interpretation, but does imply that certain favoured interpretations will be linked to specific contexts.

At the simplest level, tools simply signify themselves as objects. Something that looks like a hammer is probably a hammer. However, even this simple point hides the equally obvious notion that many objects can be used for hammering, and so the concept of 'hammer' is part object and part function. In other words, something that looks like a hammer can be used as a hammer. This suggests that objects signify both form and function. I also propose that each type of signifier can hold multiple perspectives.

Product semantics

An area of interest in the field of industrial design is known as 'product semantics'. This proposes that, rather than simply reflecting a specific function, an object communicates meanings with its prospective users. It is assumed that, in assigning a 'meaning' to a product, one will draw upon a whole set of referents, e.g., through analogy, metaphor, allusion, culture, etc. to interpret a product and to provide clues as to the product's interpretation. A popular television programme used to ask panellists to determine what a particular object was for. Often the objects were tools from industries that had been long obsolete, and the panellists would inspect the objects to look for handles, shafts, blades, etc. that could be used to develop an explanation. Their interpretation of the object depended both on 'seeing' the components of the object and on being to draw analogies with similar components and objects.

Product semantics assumes that users can read or interpret objects in terms of their meanings. It would seem that the idea of reading suggests a strong cognitive engagement, in which the person contemplates the object in order to extract the meaning. However, this meditation upon an object's meaning does not seem to be an everyday practice. Rather, we are more likely to respond to the immediately apparent features. In this way, the appearance of a tool will cue its meaning, e.g., in terms of how to pick it up, how to use it and what to use it for. An interesting study is to take an everyday object and place it on the table in front of you.

First, facing the object, close your eyes and then open for 2 seconds and close them again. This fleeting glimpse of the object will help show which features of the object are dominant and most conspicuous (they are not always the features that you might predict). The conspicuous features could be those that inform the initial, rapid interpretation that we might make in our everyday interactions with objects. Ideally, the role of a designer is to ensure that the conspicuous features are those that point to appropriate meanings.

Signifying form

The shape of a tool's handle invites or affords or supports a particular type of grip. Many handles can support more than one type of grip, e.g., a screw-driver can be held in a precision grip when placing the tip of the screwdriver into the head of the screw, and can be held in a power grip when driving the screw; a spoon can be held to drink soup or eat pie. The handle offers a range of grips and also a range of postures for using the tool. Selection of the grip is a function of previous experience, e.g., the needle holder shown in Figure 5.1 might look like a pair of scissors and people might mistakenly attempt to use a grip that is appropriate for scissors when using these. Selection of the grip is also a function of the constraints under which one is performing the task, e.g., eating in a restaurant vs eating at home, eating soup vs eating pie, etc. The selected grip can be related to the morphology and motor skills of the user, e.g., a young child may well grip the handle in a manner different from an adult due to the differences in hand size and fine motor skills. At this level, the meaning of the tool relates to its interaction with the specific person who is using it. But even this discussion implies that 'meaning' becomes intertwined with a broad range of contextual features that shape our interpretation both of the tool itself and the manner in which it is to be used. Defining the appropriate manner in which to respond to the form of the tool then becomes a matter of making judgements based on multiple sources of information.

Aesthetics

It is worth noting that the root of the word 'aesthetics' (Gr. *aisthetika*) refers to that which is perceptible through the senses, not simply visual but also tactile. What one is most likely to sense, in terms of aesthetics, is some notion of the value of the tool. The appearance of the tool can influence a whole host of possible value judgements. For instance, is the tool well made and expensive or does it look cheap? This is not simply a matter of the choice of material used to make the tool (although this is clearly a major factor), but also the shape and appearance of the tool. Tools can also be considered in terms of the prospective user group, e.g., tools that have been made or modified for use by people with physical impairments will look different from other tools, perhaps because they have larger handles. Finally, the general pattern of wear on tools can be used to signify their value, e.g., contrast the appearance of a bright new hammer that might be in the toolbox of the unenthusiastic DIYer with the worn and dirty hammer of the practicing builder. Thus, one can look at well-worn tools and imagine using them. Equally, one can look at well-crafted tools and imagine how one might use them. Tools can possess aesthetic value in that their form is attractive, or that they look useful.

One can make such judgements of the feel of the tool, particularly in terms of how the tool feels in your hand and what it is like to use. Consider writing with a well-crafted fountain pen on beautifully smooth writing paper; the pen seems to glide across the surface of the paper and the hand merely guides it, i.e., there is no sense of 'gripping' the pen. In this manner, the morphological engagement between writer and pen seems to disappear and one is only aware of the words flowing onto the paper. A similar relationship often arises when experienced tool users work with familiar tools. In this respect, the meaning of the tool is as much to do with the emotional response of the user to the object, as to the tool itself.

Signifying function

For many tools, the form of the tool seems to imply a rather pedestrian description of a particular function. It is clear what one is supposed to do with the tool, with the handle being clearly separated from the rest of the tool. Thus, the tool is designed to signify its function. However, the intended function of the tool reflects a whole host of cultural expectations. The designs of spoons reflect attitudes to eating as much as the function of holding food. The designs of saws reflect notions of appropriate ways of cutting wood. In terms of saw design, a great deal of effort has been put into defining the size, shape and angle of the teeth of the blade to ensure an optimal cut. The dimensions of the saw are such that the cutting stroke does not exceed the reach of the average user, the handle of the saw is designed for a particular range of hand sizes, etc. All of these design decisions are taken to reflect a particular manner in which the saw is intended to be used. In other words, the design of the saw represents the theory of use of its designers. Such a 'theory' encompasses the grip and posture that the user is supposed to adopt, the type of cutting action that is assumed to be employed with the saw, the sort of materials to be sawn, etc.

Another way to consider this point is to say that tools reflect certain gestures of action: we are able to 'pantomime' the use of tools in fairly consistent manners, which implies a set of stereotypical expectations as to how one is supposed to act with tools. The design of tools may reflect these stereotypical functions. The fact that one does not receive an instruction booklet with a saw suggests that these assumptions are widely accepted and the knowledge of 'how to saw' is widely held. However, I am not sure that these assumptions are true.

As we discussed in earlier chapters, the notion of affordance implies a relationship between object and use. The design of tools certainly implies an affordance for grip, but not necessarily an affordance for action. For example, in a laboratory study I asked people to pick up a screwdriver under different conditions. The conditions were simply to pick up and hold the screwdriver, to pick up the screwdriver in order to drive a horizontally fixed screw or to pick up the screwdriver in order to drive a vertically fixed screw. I measured (using a camera-based movement tracking system) the distance between thumb and forefinger (to evaluate grip width) and the orientation of the hand. It is known that grip width varies according to the type of object being picked up and also that hand posture varies in response to the action to be performed (see Chapter 11). I found little evidence of preparatory modifications to grip width or hand posture in this study. In other words, the participants tended to pick up the screwdriver in much the same manner in all conditions and *then* to modify their hold of the screwdriver to perform the tasks. It is perhaps noteworthy that the grip used tended to reflect a power rather than precision grip in all conditions, and that, for the screwdriving tasks, the participants would adjust to a precision grip to home the tip onto the screw heads and then revert to power grip for driving the screw. While this might appear relatively obvious, it does raise an important question as to what action the handles of these tools are designed to support: it could be proposed that the handles are designed with the intention of supporting a specific type of grip, rather than supporting a specific type of action.

Signifying operation

The preceding sections have focused on the meaning of tools as static objects, i.e., as things that are lying on the table. However, as soon as you pick up a tool, a new set of meanings emerges. For instance, the balance of the tool in the hand can set up expectations of the posture that will be needed to hold the tool. Thus, a power drill might be heavy and pull the wrist forwards, which means that to use it might require two hands or might require you to keep the wrist straight through pulling the drill up. The appearance of an angled handle for pliers and hammers (see Chapter 6) might look a little unusual, but when you pick up a tool designed with an angled handle, it sits quite differently in the hand and suggests a different style of use. In this manner, the angled handle signifies a pattern of operation different from what might be appropriate for a straight handle.

Some handles are designed with indentations for the fingers. Such a design has already been argued to be very poor ergonomics (see Chapter 6), but it does imply a specific notion of use that the designer had in mind when creating the tool (as well as a single dimension of the human hand, which is one reason why such designs are so poor).

Tools as 'objects to think with'

Levi-Strauss [1] viewed tools as 'objects to think with'. This implies that an additional function of a tool is to reflect the knowledge necessary to perform a task. Thus, when a surgeon selects a particular instrument, this selection reflects a whole set of knowledge regarding the particular stage of the particular operation that is being performed. It is plausible that the tool reflects a 'short-hand' description of the procedure and that this also provides some cognitive support for the activity. For instance, a particular procedure might involve the use of four instruments in sequence, and the surgeon might know that when the third instrument has been put down it is time to pause and check the state of the operation. Alternatively, when I was observing jewellers, it certainly felt as if actions such as filing metal were performed rhythmically, e.g., make a dozen strokes with a file, pause to blow off the filings and repeat until four sets of these actions have been performed and then visually inspect the piece. Of course, such heuristics are very much a matter of individual practice, but the very plausibility of the description implies that the use of the tool is embedded within a space of assumptions, experiences and tacit knowledge to such an extent that the very act of using the tool provides a means of marking and traversing this space.

Furthermore, selection of a tool represents a particular approach to performing a task. For instance, my approach to cut meat will differ when I select a meat cleaver to when I select a carving knife. The selection of the tool both reflects my intentions and also influences these intentions through constraining my actions. If I decide to modify my behaviour, I might need to change the manner in which the tool is used or select an alternative tool.

Cultural significations

As we saw in Chapter 4, stone-tool manufacture seemed to persist in a similar fashion for millennia. This might reflect some limitation in the evolution of the people making the tools, but it seems somewhat odd that so little change would occur over such a long time period. An alternative perspective is that the tool making reflects a tradition of manufacture, i.e., there was a culturally accepted view of what the tools should look like and the manufacturer was directed at producing objects to meet this view. In a like manner, the design of hammers has remained relatively unchanged since the Bronze Age. When tools do change, this might arise from the evolutionary drivers proposed in Chapter 6. Tools, therefore, might reflect adaptations to new ways of working and, in turn, offer yet newer ways of working.

Sometimes tools became stylized and associated with ritual activities. In this manner, the tools began to stand for a particular set of ceremonial actions and could be represented as a figurative design or could be inscribed to carry words and symbols to increase the ceremonial power of the object. In modern times, 'tools' can often function as symbols. Thus, the hammer and sickle of the Soviet Union symbolized peasants and workers in unity. Indeed, the very word 'tool' is often used to symbolize the general notion of something to assist in performing a task. On computer systems, the icons used to denote functions such as 'cut' or 'paste' are images of scissors and brush. In these cases, the image of a physical tool is used to denote a sequence of actions that in pre-computer times were performed using these tools for some aspect of the task.

Physical tools/cognitive tools

This book has focused on tools that support physical activity. The idea of the tool is as an object that can augment physical activity, e.g., by harnessing power or by making movement more precise. However, we are surrounded by tools that support cognitive activity. For example, a clock displays a measure of an abstract division of the day into units of hours, minutes and seconds. Thus, the clock functions as a tool to measure the passage of time at a greater level of precision than one can normally achieve. The issue of cognitive tools will be discussed further in Chapter 9.

Discussion

In this Chapter, I have considered some of the factors that might influence our ability to interpret the meaning of tools. My point is that even the most mundane of tools have the possibility of carrying some meaning to the user. Perhaps this meaning is simply in terms of what the tool can be used for, alternatively the meaning might relate to some notion of the tools 'aesthetics'. Thus, looking at a teaspoon in the glass case of a museum, perhaps as part of an exhibition of Georgian artefacts, might provoke different reactions to looking at a teaspoon in a restaurant, which might, in turn, differ from looking at a 'tea stirrer' in a fast-food outlet. The function of the three artefacts remains the same, but the form will differ, as will the value that one places upon them. For instance, the tea stirrer is clearly designed for immediate use and is to be disposed of when it is finished with, but the teaspoon is designed to be washed and reused.

By having a chapter devoted to the meaning of tools, I am trying to point out that these everyday objects can be read and interpreted in terms of different meanings, and that we respond to different meanings depending on the contexts in which we use the tools. One implication for this is that tool design is as much a matter of relating the appearance of an object to its potential context of use as it is to the form and function of the tool. Indeed, a beautifully engraved sterling silver teaspoon might look distinctly out of place on the workbench of a mechanic but might appear perfectly at home in the sitting room of a Stately Home.

8 How tool use breaks down

Errors in what we do can give us insight into how what we do is planned and controlled.

(Smyth *et al.*, 1987)

Introduction

In this chapter, the focus of attention will be on the ways in which the practice of using tools can breakdown. In broad terms, I am considering 'breakdown' from two perspectives: on the one hand, breakdown can simply mean that an action has not gone as planned, e.g., in terms of human error; on the other hand, breakdown can mean that the ability to perform the action is missing, e.g., in terms of physical or neurological impairment. Consequently, the chapter will be divided in two parts. The first will examine human error, with particular relevance to slips and other forms of error. This will also involve consideration of accidents and injuries arising from tool use. The second section will focus on physical and neurological impairment, with particular emphasis on 'apraxia', i.e., failure in the ability to perform coordinated, object-based actions. This will also involve consideration of the possible regions of the brain in which tool-related information is stored.

Human error

Contemporary theories of human error emphasize that people form part of a 'system' and that the failure of an individual often represents failure throughout the system [1]. Thus, many industrial accidents can be related not only to an individual making a mistake but also to management, technology and work practices that 'allow' such a mistake. It is possible to consider tool use from a systems perspective in order to determine what errors might arise. One approach to such a perspective is to follow the forms of engagement argument proposed in this book. Let us take the simple example of tightening a screw that has come loose on the hinge of a cupboard door. Table 8.1 shows some of the ways in which this action could fail.

Table 8.1 Errors in a simple task

Forms of engagement	Failure mode	Cause	Consequence
Environmental	Poor access to screw head	Door obscures access	Constrained posture
		Screw head is worn	Difficult to gain purchase
Morphological	Poor grasp of screwdriver	Poor grip arising from slippery handle or weakness	Low torque
Motor	Fail to exert sufficient torque	Weakness, poor purchase, poor grip	Screw won't turn
	Fail to fit tip into screw	Wrong tip, lack of precision in fitting	Tip slips from screw head
Emotional	Fail to produce 'proper' finish	Screw not flush with hinge	Unsightly appearance, door not closing properly
Perceptual	Misinterpret type of screw head	Screw differs from those used on the other hinges	Use wrong screwdriver or fail to fit tip to head correctly
Cognitive	Select different tool	Use knife to hand rather than go to toolbox	Risk of injury, poor purchase
Cultural	Turn screw the wrong way	Expect clockwise screw thread	Screw does not loosen

Table 8.1 suggests that even a simple task can fail in a variety of ways. It is worth noting that many of the failures interconnect. For example, if I select a different tool, such as a knife that is lying on the kitchen work-surface rather than go to my toolbox, I run the risk of poor purchase (i.e., the tip of the knife will not properly fit the screw head), which could lead to either failing to turn the screw or the knife slipping.

Whereas ergonomics talks of human error, cognitive psychology has tended to employ the word 'slips'. Slips have provided a great deal of evidence for theories in psychology. For example, the study of slips suggests different stages in the planning and production of speech, typing and hand-writing. Many of the descriptions of human performance that have been derived from slips are hierarchical in nature, i.e., they assume a sequence of processes from specification of intention to action. There has been little, if any, research that has focused directly on the use of tools, so the following will necessarily be somewhat speculative.

In a review of 'cognitive skills', Colley [2] suggested that theories tend to follow a similar structure. A 'goal' is specified, which is then decomposed into constituent parts that are then combined into an appropriate response.

Table 8.2 Slips in cognitive skills

Typing	Handwriting	Speech
Input from paper	Define grapheme, or word	Define message
Parsing of letters into string	Define allograph, letter shape	Functional content
Translation of letters to key	Define movement to create graph	Positioning of words
Execution of key press	Execution of movement	Phonetic programme of words
		Articulation

From Table 8.2, we can see that typing, handwriting and speaking exhibit similar patterns. Much of this work arose from the study of errors that people make when performing these tasks.

From the discussion of tool use in this and previous chapters, a similar description can be put forward. From Table 8.1, it is possible to think of failure in terms of selecting an inappropriate tool, problems of grasping and manipulating the tool or poor interfacing between the tool and objects in the world. The implication of this (following the previous ways in which slips have been considered) is that tool use requires specification of a tool and its appropriateness, a specification for grasping and manipulating the tool, and a specification of how the tool will interface with objects in the world. The ordering of these specifications is not straightforward; the user could first specify how the tool will interface with objects in the world, e.g., by determining the type of screw head, and then use this to specify a tool, i.e., screwdriver with an appropriate type of tip, or the user could have a looser specification of the interfacing, e.g., determining that a nut is hexagonal and of less than 1 cm diameter, and then bringing a collection of spanners to test on the nut. In the first instance, the 'specification' of the tool is a direct consequence of interpreting objects in the world, and in the second instance, the 'specification' will be created during practice. It is likely that these instances reflect variations in activity, i.e., if you are changing the tyre on a bicycle and have already used a particular spanner to remove the wheel, then you will seek the same spanner on subsequent occasions. Indeed, bicycle manufacturers often supply a multi-purpose spanner with a variety of heads for the different standard-size nuts that are used.

Following from Chapter 5, one could suggest that the first instance represents a form of workmanship of certainty, in which a clear link can be made between object and tool, and the second instance represents a form of workmanship of risk, in which the link is somewhat ambiguous and needs to be clarified through practice. In either case, failure to establish the link between tool and object can lead to failure of performance, e.g., the wrong tool could be selected leading to poor interfacing between tool and object leading to slippage.

Table 8.3 Relating hierarchical control from cognitive skills to tool use

Process	Activity
Define	Specific goal
Input	Respond to features of tool and environment
Parsing	Specific appropriate posture and grip
Action	Specific appropriate action
Execution	Perform action

A second form of slip is between handle and hand. This could arise simply from failure to grip the handle firmly, e.g., due to musculoskeletal problems, or from the handle slipping from the hand, e.g., due to the handle becoming slippery. These failures imply problems in either the initial grasping of the tool or in the moment-to-moment coordination and control of the tool. These slips could also arise from using an inappropriate grip, e.g., using a power grip when a precision grip is required, which might, in turn, be influenced by the posture adopted when using the tool. This latter point implies an association between knowing how to use a tool and particular types of failure.

The foregoing discussion has expanded the aspects of tool use that require specification, i.e., specify tool (based on objective and interface between tool and object), specify posture, specify grip and specify 'goal'. Having said this, it is clear that we do not consciously involve ourselves in such specifications every time we use a tool; we simply 'know' which tool to use. However, the point to note is that one can propose a rudimentary description of using tools, based on possible failure modes, which implies a set of specifications (see Table 8.3). In the next section, we investigate reports of injuries and accidents associated with the use of hand tools in order to consider whether there is evidence for failures of these different specifications.

Accidents and injury when using tools

Accidents involving tools are typically considered in terms of injury. Before reviewing some of the data relating to this aspect of accidents, it is worth pausing to consider other consequences of accident and error. Returning to Table 8.1, it is probable that many of the errors will not lead to injury but will lead to either frustration or other ways in which the task is thwarted. Thus, error can lead to delays in working or to poor quality finish on work, as well as to injury.

Around 8 per cent of US workplace compensation claims involve accidents relating to hand tools [3]. Similar statistics seem to hold for most industrialized countries, e.g., figures from Australia and the United Kingdom are of a similar order. In addition to compensation costs, there are a number of other costs associated with such accidents, as the following example from a UK Health and Safety Executive booklet illustrates, 'An operative [on a North Sea Oil platform] hit his hand with a 7 lb hammer – cost £2200. This high figure resulted from the high costs of obtaining treatment and underlines the variations in costs that can occur depending on circumstances of particular cases' [4].

In a survey of woodworkers in the United States [5], around two-third of respondents reported having some sort of injury (indeed, 5 per cent had suffered partial amputations). The most commonly reported tools were hammers, chisels, drill presses and table saws, and they had the highest injury rates (between 3 and 5 injuries per 1,000 person-hours of use). In the home, accidents arising from the use of hand tools are also quite common. For example, the DTI publishes Home Accident Surveillance Surveys (HASS), based on interviews with patients in Accident and Emergency units in hospitals. The responses are then used to estimate national figures for a given year. I have extracted the data shown in Table 8.4 from the 1999 HASS report [6].

Thus, many accidents and injuries are associated with tools that either have sharp edges or which require force to operate them. The obvious implication is that injuries arise from either slips of the tool, in which the sharp edge cuts or otherwise injures the user, or overexertion, in which the user becomes injured as a result of either impact from the tool or from the strain required to continually operate the tool.

A detailed investigation of the relationship between injury and tool use was reported by Aghazadeh and Mital in 1987 [7]. From the survey, the hand tools that were used in the majority of reported accidents were knives (44.3 per cent), with hammers, wrenches and shovels reporting between 5 and 10 per cent of accidents. For power tools, the most commonly reported were saws (33.6 per cent) and drills (17.6 per cent), with grinders and power hammers accounting for some 8 per cent of accidents. The vast majority of the accidents fell into two classes: (i) around 66 per cent of the accidents involved being struck by or against the tool (including hitting, cutting, etc.) and (ii) 25 per cent of the accidents relate to overexertion. Table 8.5 classifies these injuries into types and body regions.

Table 8.4 Accidents in the home arising from the use of 'hand tools' (the numbers in brackets give figures from HASS, 1999)

1,000–3,000	3,000–5,000	5,000–10,000	10,000+
Food slicer (1,244)	Mallet (4,353)	Saw (6,145)	Kitchen knife (28,020)
Spanner (1,500)	Screwdriver (3,823)	Scissors (6,219)	
Tin opener (1,536)			
Cutlery (1,774)			
Pen (1,975)			
Plane/sander (2,268)			

Table 8.5 Relating injury type and region to type of tool

Injury	Hand tool (%)	Power tool (%)
Type		
Cuts, lacerations	61	53
Sprain, strain	19	25
Contusion, bruising	10	8
Region		
Upper extremities, including hands	59.3	51
Back	16.7	15.4
Lower extremities	9	22.1

One American insurance company presents some simple safety procedures that relate to the use of hand tools [8].

1   Use the right tool for the job. Do not use a wrench as a hammer, or a screwdriver as a chisel. Get the right tool in the right size for the job.
2   Do not use broken or damaged tools, dull cutting tools or screwdrivers with worn tips.
3   Cut in a direction away from your body.
4   Make sure your grip and footing are secure when using large tools.

The advice relates both the types of accidents and injuries reviewed here and also to some of the proposed forms of error presented earlier. In particular, points 1 and 2 relate to tool selection and imply that people might be motivated to use an inappropriate tool in order to save time and effort in fetching the correct tool; points 3 and 4 relate to the actions required to use tools and imply that people might use inappropriate posture or grip when using tools. From this, one can propose that accidents relating to hand tools often relate to problems of morphological and motor engagement, e.g., in terms of the tool slipping from the hand either due to problems in the design of the tools or due to mishandling of the tool, and from problems of cognitive engagement, such as selecting the wrong tool. Tool selection would appear to be partly influenced by interpretations of task demands and partly by other demands, such as the trade-off between going to fetch the correct tool and 'making do' with the tools to hand. This latter point suggests that people have a flexible interpretation of what constitutes an appropriate tool, even if this means a somewhat risky activity in using the wrong tool.

The preceding discussion suggests that most accidents relating to hand tools lead to physical injury, such as cuts, etc. The implication is that the tool 'slips' during use, and that this results in injury to the user. However, it is worth noting that hand tools can also lead to injuries to the face (especially the eyes), e.g., from chips of wood flying from the object being worked, and to other parts of the body, e.g., from tools being dropped or released during use.

Overexertion was identified as a significant source of injury, and a major concern to ergonomists is the potential for prolonged use of hand tools to cause occupational injury. Such injuries are most often associated with the use of power tools, and can also be linked to non-power tools. Shock loading, e.g., is particularly common when hammers are used incorrectly, and can lead to increases in muscle tension. For power tools, the incidence of Stenosing Flexor Tenosynovitis (popularly known as 'trigger finger') has been shown to have a strong correlation with the use of hand tools [9]. There are several possible causes for trigger finger, but the most plausible relate the vibration of a power tool to either vibration passed to the skeletal structure in the hand holding the tool, such that one can feel after-effects of the vibration, or to the fact that certain frequencies lead to vascular problems and so can cause longer-term problems relating to the supply of blood to muscles and tendons. People with 'trigger finger' (and vibration-related symptoms) tend to exhibit reduced blood flow in the hands associated with the symptoms which, in turn, results in lower temperature of these regions. The UK Health and Safety Executive advises that power tools be fitted with a vibration absorbing grip and, where possible, avoid tools that transmit vibration to the hands [10]. These, and associated design issues, were discussed in detail in Chapter 6. In general, injuries directly arising from the use of hand tools appear to be due primarily to force and posture. Some tools either transmit forces to the user or require excessive force to be applied by the user, and many of these problems can be reduced through adequate design of the tools. A common complaint in manufacturing industries is carpal tunnel syndrome. In this condition, the median nerve in the wrist becomes injured as it passes through the carpal tunnel. The condition is often associated with highly repetitive work having high force application, e.g., pressing fixtures into place using the fingers or thumbs.

The brief discussion of errors, accidents and injuries relating to tools suggests that tool use can breakdown in a number of ways. It is proposed that the notion of forms of engagement provides a convenient way to characterize these breakdowns, as shown in Table 8.6.

From the review of human error, one can begin to sketch an outline theory of tool use in that there are three clear sources of failure, i.e., grip, posture and tool. From the accident and injury data, one can postulate that each of these sources involves failure of specification, i.e., tool would require {specify function of tool, specify action to use tool, specify object to act upon}, grip would require {specify posture to use tool, specify intention in using tool, specify action to use tool}, posture requires {specify action to use tool}. Failure of any of these specifications could lead to some form of error or accident. Of course, this rather assumes that tool use is static, i.e., that the mind 'writes' a specification and the body does the rest, and pays little

Table 8.6 Relating errors, accidents and injuries to forms of engagement

Forms of engagement	Error	Accident	Injury
Environmental		Slip of tool on object	Cuts, hits and other injuries from tool or from object and waste Vibration from tool or through impact
Morphological	Select wrong grip	Slip of tool in hand	Grip posture
Motor	Adopt wrong posture	Slip in action	Musculoskeletal strain
Emotional			Stress
Perceptual	Select wrong tool		
Cognitive	Specify wrong tool		
Cultural			

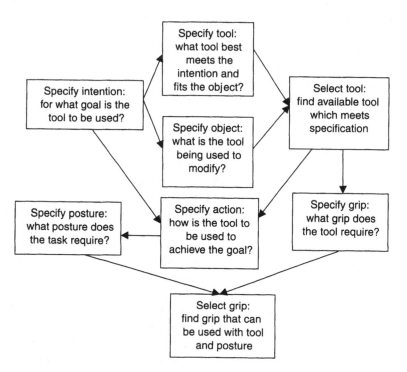

Figure 8.1 Relating aspects of tool use that can lead to errors, accidents and injuries.

attention to using feedback from tool or environment. As we saw in Chapter 5, a hallmark of the skilled user of tools is the ability to assimilate such feedback in order to modify and correct activity. Having said this, Figure 8.1 provides an outline of tool use. The idea underlying Figure 8.1

is that tool-using actions can fail either through omission of one of the boxes, or through failure to correctly perform the action in the box. Thus, in the 'select tool' box, the person might opt for a tool that is readily to hand rather than leaving the job and going to fetch the correct tool. The basic assumption is that, even without considering how one actually uses the tool, there are several ways in which tool use can breakdown.

Tool use and motor impairment

Tool use can be affected by impairment of motor and morphological engagement. These impairments can range from a weakening or other loss in the ability to grasp objects that would limit morphological engagement, e.g., through arthritis, to loss of coordination in movement which would limit motor engagement, e.g., through Parkinsonian tremor. There is a great deal of work spent on developing tools and aids that can support people with such impairments, and much of this work is finding its way into the design and marketing of special-purpose products, or means to modify standard products. Thus, for sufferers of arthritis, it is possible to modify handles to reduce the amount of grip force required, perhaps through the use of large, foam handles which can not only support a weakened grip but can also provide cushioning against impacts arising from the use of the tool.

In addition to impairments arising from physical problems, it is also possible for the environment to restrict the users' abilities to handle tools. Thus, for instance, people working in hazardous environments might need to wear gloves that can influence morphological engagement. Thus, the interaction between glove design and tool design becomes particularly important. For instance, gloves can alter the dimensions and shape of the hand (and so limit morphological engagement), restrict movement (and so limit motor engagement) and reduce the ability to sense feedback through the hand from the tool (and so limit perceptual engagement).

Commercial divers and astronauts will need to use tools in environments that do not support 'normal' forms of movement. For divers, problems can relate not only to the changes in the dynamics of movement of objects under water, but also the ways in which poor visibility can affect perception and working with breathing apparatus can affect cognition. Divers cannot simply exert more muscular force to move objects underwater, but need to develop strategies for combining the movement of the objects with their own movements. Astronauts face similar problems in that the objects they are manipulating do not obey the laws of physics that govern land-based objects.

In the next section, we turn our attention to the ways in which the representation of tool use in the brain can become impaired, and how this might affect performance. The main focus of the following discussion will be on the class of disorders known as apraxia.

Apraxia

In the field of neuropsychology, a set of conditions has been observed that lead to patients being unable to perform particular actions. Early in the twentieth century, Hugo Liepmann identified these conditions as 'apraxia', i.e., a failing of praxis, or action movement. In other words, patients with apraxia lose the ability to perform the sort of skilled actions that we take for granted in our everyday lives. The loss of ability is not related to any weakening, which implies that the deficits relate to higher-order cognition rather than physical impairment. Patients with apraxia have difficulty in manipulating everyday objects and performing everyday actions. Liepmann proposed that there were two basic kinds of apraxia: ideational, which relates to a breakdown in the mental representation of motor actions, e.g., being unable to act out (pantomime) using an object; and ideomotor, which relates to a breakdown in the translation of mental representation of action to the motor commands needed to perform the action, e.g., an inability to copy someone else's movements or gestures. Over the years, other forms of apraxia have been proposed but this chapter will focus on those forms of apraxia that are linked with the use of tools, e.g., inability to produce tool-mediated action, either through gesture or action sequence. (See Figure 8.2.)

Ideomotor apraxia

Patients with damage to the parietal area of the left hemisphere, perhaps as the result of a stroke, Alzheimer's disease or a tumour, can sometimes show errors in 'pantomiming' (or acting out) the use of a tool. Thus, patients might be unable to gesture the action of cutting a slice of bread from a loaf, because they are not able to make the right arm movements [11].

Apraxic patients can name objects (when presented with pictures). This means that the patients are able to recognize the object, in terms of what it

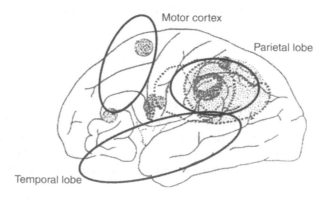

Figure 8.2 Approximate distribution of regions of the left side of the brain in which lesions are associated with apraxia. Lesions are shown as grey areas.

is called or in terms of specific attributes of the object. However, when asked to respond to the function of the object, their ability is impaired. For instance, in one study [12], Alzheimer patients were shown line drawings of common objects, such as a rowing boat, telephone, saw, etc. Part of the study involved pantomiming object use, and 72 errors (out of 160 elicitations) were recorded. The pantomime production errors fell into 8 broad categories (the numbers in parentheses indicate the number of errors in each category):

1   Complete but minor error (27) 'drum': hold sticks, hum tune, but not move hands;
2   No response (12);
3   Incomplete (10) 'apple': chewing but not holding or biting apple;
4   Vague (7) 'umbrella': raising hand in the air;
5   Complete error (7) 'telephone': typing;
6   Correct but not stereotypical (4) 'gun': load gun rather than aim and shoot;
7   Body part as object (3) 'pen': writing with index finger rather than holding a pen;
8   Perseveration of previous response (2).

One interpretation of this study is that the difficulties in pantomiming object use is related to problems in accessing representations of the objects' function. However, it is worth noting that these patients had reasonable performance on a naming task and that many of the pantomime errors have some correct components, i.e., categories 1, 3, 4, 6 and 7. My reading of these data is that some 44 of the actions contained a 'correct' part, albeit either incomplete or atypical, and 28 were completely erroneous. This implies that the breakdown is not total, but rather that the patients were accessing only part of the action representation related to specific objects.

   Other observations describe errors in orienting the hand around object, and using a body part as the tool (e.g., miming the action for 'hammer' using the hand rather than holding a hammer) or using the body part as object (e.g., pretending to hammer onto the hand). Thus, in another case, a patient was asked to use everyday objects, such as a spoon. She was not able to form the grasp or manipulate the object correctly, nor could she recognize or describe appropriate hand gestures for using objects for different tasks [13]. This suggests an inability to relate movement to specific goals. In such cases, the patients lose the ability to access representations of coordinated sequences of movement. This does not necessarily mean that the representations are missing; it is possible for patients with apraxia to repeat stereotyped movement sequences even when they do not intend to, e.g., to perform a reaching action to a cup even when instructed not to – in such instances, the presence of the cup serves as a cue for the movement sequence, which can be performed against the person's will. This suggests that apraxic patients are able to perform movements that are embedded in familiar contexts; thus, 'having a drink' would involve reaching for a cup and raising to one's mouth. When the actions are removed from this context, e.g., raise a glass but do not drink, or pretend to raise the glass, then problems arise. Thus, ideomotor apraxia would appear to stem not from the loss of the representations of movement, so much as the failure to coordinate and access these representations. Indeed, in a study comparing hand–arm movement for simple gestures, such as raise a spoon to one's mouth or salute, it has been shown that apraxic patients tend to follow similar arm movements and only modify hand movement, which leads to incorrect responses [14], which suggests that they are employing an incomplete movement sequence.

Problems associated with sequencing of actions are also related to frontal lobe damage, possibly leading to disruption of action programmes. In a classic account, Luria tells of a patient who, '... when asked to light a candle struck a match correctly but instead of putting it to the candle which he held in his hand, he put the candle in his mouth and started to smoke it like a cigarette. The new and relatively unstabilised action was thus replaced by a more firmly established inert stereotype' [15].

Conceptual and ideational apraxia

Conceptual apraxia results in problems in knowing which object to use, i.e., errors in associating tool with action, or inappropriate selection of tool for task. In a related condition, ideational apraxia relates to problems in recalling all the steps in a tool-using procedure. This tends to describe the inability to follow any form of planned action.

Patients with ideational apraxia might fail on pantomiming tasks, but can accurately and appropriately use an object when it is presented to them [16]. For instance, a patient might be unable to act out the sequence 'pour a drink from a jug into a cup', but, when confronted with jug and cup, could easily perform the actions. Another set of symptoms relates to the ability to sequence tasks properly. In other words, an apraxic patient could easily misorder the tasks 'turn on gas', 'light match to ignite stove' and 'blow out match' to near fatal consequences. Such problems imply that this form of apraxia is less about being unable to perform the actions, and more from the inability to access the concepts or ideas that underly both the actions and their sequencing. A common means of demonstrating this is to ask ideational apraxic patients to place pictures of steps in a task sequence in order; while this is a trivial task for most people, patients with apraxia are unable to produce correct order of tasks. Having said this, there are reports of patients who maintain an ability to sequence photographs, but are unable to perform the task sequences in everyday life.

Neglect

Before concluding discussion of neurological deficits, it is worth considering some aspects of 'neglect'. The term neglect describes a condition in which the patient is not able to attend to one side of their body, either ignoring objects on that side or being unable to move within the space on that side of their body. Typically, neglect patients do not have impairment of motion, i.e., they *could* move on that side of their body, but do not. Often neglect occurs on the left side of the body, following lesion to the right hemisphere. A simple test of left-sided neglect is to present patients with a sheet of paper instructing them to cross out specific features. For example, a page of 0 and X could be presented, with the instruction to cross out all of the 0s. A neglect patient will cross out most of the 0s, but leave some on the left side of the paper. At one level, this might reflect some problems in attention, with the patient being unable to attend to specific regions in space.

In some recent studies at Birmingham, it has demonstrated that providing a neglect patient with a means of extending space, i.e., a stick held in the hand, could improve some aspects of performance [17]. In other words, the stick appeared to function as a means of extending the region in which the person operated and the range of space that could be attended. This is interesting in that it revisits the earlier proposals (see Chapter 1) that tools were extensions of the user; in the case of these neglect studies, it would appear that the tool and the hand were sufficiently integrated to improve visual awareness of space.

Conclusions

From this discussion, we can assume that the brain stores information about objects and their associated actions in specific representations. From studies into brain activity in monkeys, it has been shown that a strong link exists between grasping an object and firing of neurons in the parietal cortex. What is interesting about this finding is that the same neurons fire when the monkey only looks at a graspable object [18].

From a different perspective, brain scans have been performed when participants are asked to name pictures. Thus, for example, more activity occurs in the premotor cortex when participants have to name tools than when they have to name animals [19]. In another study [20], it was shown that activity in both the left premotor and left parietal cortices was higher when participants were presented with images of tools than for images of faces, animals or buildings. Furthermore, the same cortices showed significantly more signal change when participants were asked to name the images of tools than for the other images.

The main problem facing patients with apraxia relates to the accessing of stored representations relating to object use. This suggests that such representations must involve knowledge of the object and how it can be used, together with knowledge of how to manipulate and use that object [21]. The former knowledge can be cued through pictures and put into words, whereas the latter would appear to be mainly accessible through movement and can be very difficult to put into words. Asking someone to perform the action in the absence of the object, e.g., 'pantomiming', therefore requires eliciting implicit knowledge relating to movement sequences. As Jeannerod states, 'The basic elements of the representation remain the same, but the context in which they are used selects different subsets among the possible configurations and triggers different strategies. The effects of lesions reveal that these operations are not limited to a particular neural system, but involve a large network distributed across systems for semantic and pragmatic processing' [22].

Most of the research reviewed so far has focused on lesions in the left hemisphere. However, a case study of a patient with right hemisphere damage demonstrated that he also experienced some problems with performing sequences of object manipulation tasks [23]. For instance, he was asked to perform the task of wet shaving; after he had used a brush to put shaving soap on his face, he then turned the brush sideways and used it to scrape off the soap, i.e., the brush was used like a razor. There are several possible explanations for this behaviour, e.g., the sequence of shaving could have been sufficiently compiled for the patient not to realize that there had been no break (in other words, he went straight from the task 'apply soap' to the task 'shave off soap'), or the feeling of the brush in his hand could have reminded him of a razor or brushing and shaving exhibit similar motor behaviours so he persevered with using the brush. Irrespective of which explanation holds, the main point to note is that the patient exhibited errors in his ability to sequence tasks. What is interesting is that his knowledge of the everyday objects and how to use them appeared to be intact (on the basis of other tests that were performed by the experimenters). Furthermore, the patient appeared unaware that he was making errors, or when he did note an error seemed unable to halt the action. Interestingly enough, when he was provided with step-by-step instructions, his error rate declined. This implied that the patient was unable to link actions into a sequence that was guided by a goal, rather than have his actions guided (or captured) by preceding actions or by other objects in the environment.

Discussion

This chapter has considered breakdown in tool use. It is often convenient, when speaking of complex psychomotor skills, to think about ways in which these skills breakdown because this provides insight into the structures and processes that are necessary for the activities. From the discussion of errors, accidents and injuries, a simple figure was constructed that suggests tool use necessitated selection of tool, action, posture and sequence of movement. When considering the consequences of errors, in terms of accidents, it was apparent that many of the accidents arise from breakdown in motor activity, e.g., slips or strain. What is not clear from the literature is whether these slips or strains arise solely from motor engagement or whether they are influenced by cognitive factors. My feeling is that the majority of the accidents will arise from slips, with the tool hitting knots in wood, slipping from the user's grasp or from the tool sliding across a slippery surface. However, it is likely that a sizeable minority of such accidents can also be attributed to the user employing the wrong tool, using the wrong grip or posture or using the wrong movements to manipulate the tool. In these cases, one can further propose that the selection of tool, grip or movement will be influenced by the user's prior experience and by environmental constraints. For example, one might attempt to use a screwdriver with a non-standard grip when a screw is particularly difficult to reach. Thus, the 'slips' could arise from a reasonable assessment of the situation and an attempt to match situational demands with the users' abilities.

From the discussion of neurological impairment, one can propose that tools require representations relating to physical appearance of the tool and the function of the tool, together with representations for the movement required to use the tool in order to perform actions, and plans governing the relationship between these movements and an overall goal. Impairments in the brain can lead to loss of access to some, or all, of these representations. Breakdown can occur either when these processes cannot be performed completely, or when the processes become captured by other features of the tasks or environment.

Taken together, these two strands of work into breakdown in tool use suggest that the tool user requires a set of representations of tools in order to be able to use them. Not only does one need to be able to recognize the features and properties of the tool by looking at it, one must also know how to physically respond to these features in order to pick up and use the tool. The brain scan studies considered in this chapter suggest that, for most adult humans and primates, the visual appearance of an object is very strongly paired with a set of specific responses, to the extent that simply seeing a picture of the object is sufficient to trigger activity in the region of the brain responsible for motor control. The forms of apraxia considered in this chapter, suggest that it is also necessary to represent the manner in which the tool is to be used, both in terms of appropriate movements and in terms of sequences of movements. The fact that neurological impairment in the parietal cortex can lead to problems in using such representations suggests that the sequences can be considered as 'plans', or at least as schema of stereotypical, well-practiced actions. This suggests that a role of cognitive engagement in the use of tools is to define the meaning of a tool and then to select an appropriate schema that can be used to govern the control and manipulation of that tool. This idea will be explored in more detail in Chapter 11.

9   Cognitive artefacts

There is little understanding of information-processing roles played by artifacts and how they interact with the information processing activities of their users.

(Norman, 1991)

Introduction

Vygotsky proposed that a great deal of cognitive activity involves the use of 'psychological tools'. By these, he meant '…language, different forms of numeration and counting, mnemotechnic techniques, algebraic symbolism, works of art, writing, schemes, diagrams, maps, blueprints, all sorts of conventional signs, etc.' [1]. The basic premise is that through such 'psychological tools', it is possible to change the structure of a given task and, by doing so, to change the cognitive operations that the task requires.

In Vygotsky's words, 'The invention and use of signs as auxiliary means of solving a given psychological problem (to remember, compare something, report, choose, and so on) is analogous to the invention and use tools in one psychological respect. The sign acts as an instrument of psychological activity in a manner analogous to the role of a tool in labor' [2]. In Chapter 7, I proposed that one could think of 'tools' in terms of their semantics, i.e., in terms of the knowledge that they can signify. In this respect, a tool functions as the signifier for the knowledge that it can be contained in a 'sign', i.e., a tool (as a physical object) can stand for a range of concepts. For example, a knife and fork could stand for implements with which to eat, but could also signify a place where one could eat, or a cutlery seller, etc. Thus, while Vygotsky is careful to distinguish between a tool as a physical artefact and a sign as a cognitive artefact, I think it can be useful to blur this distinction. A study by Hutchins [3] provides us with a good illustration of this point.

Hutchins spent a great deal of time on board US Navy ships, observing the work of the crew, particularly in terms of navigation. One problem that confronted the people involved in navigation was how best to calculate the ship's speed, based on its movement between two points that are being plotted. There are several ways in which this task can be performed, and Hutchins considers the following:

a   The navigator knows that the equation $D = RT$ and the algebraic rules for transforming this equation, together with the facts that there are 60 minutes in an hour and there are 2,000 yards in a nautical mile. From this knowledge, it is possible to solve the equation for the rate (R) in terms of the time elapsed and the distance between two points, using a pencil and paper to perform the calculations.

b   Task (a) can be performed, but by using an electronic calculator rather than pencil and paper.

c   The navigator can use a three-scale nomogram. This consists of a scale showing time, a scale showing distance and a scale showing rate. The user simply marks the known values of time and distance, and joins them with a straight line. The line intersects the rate scale to provide the answer.

d   The navigator knows that 3 minutes is 1/20 of an hour, and that 100 yards is 1/20 of a nautical mile. Thus, the number of yards travelled in 3 minutes corresponds to the number of miles travelled in 1 hour, i.e., if the ship travels 1,000 yards in 3 minutes, its speed is 10 nautical miles per hour.

Each of these approaches to solving the problem employs a combination of artefacts. The main function of the artefacts, in this example, is to structure the problem space in order to produce a different sort of problem. The person performing the task then works on the representation of the problem that the artefacts provide. This is not to say that the conditions do not employ that same 'knowledge' about ship's speed; rather, such knowledge is represented in different ways. For instance, in condition (a) the navigator will calculate the relationship between D and T in order to arrive at R, whereas in condition (c) the navigator will align the scales for D and T in order to produce R. In this instance, the use of the artefact modifies the nature of the task in at least three ways: (i) the explicit act of performing arithmetic is removed and replaced with the act of aligning two scales, which makes the arithmetic implicit; (ii) the certainty of the outcome of the task is shifted from 'have I performed all stages in the calculation correctly?' to 'have I aligned the scale on the correct values?' which makes the checking of the task explicit; (iii) the speed with which the task can be performed is significantly reduced in condition (c).

Returning to Vygotsky, we can see that psychological tools are artificial formations, in that they have been created with the specific intention of supporting particular cognitive activities, and that they are cultural products, in that they reflect accepted ways of working. To this, we can add the notion of allocation of function, raised in Chapter 5, in which the artefacts are allocated some of the cognitive activity from the task. The design of the artefacts reflect a set of culturally accepted assumptions about what the tasks entail and how best to represent the problem. In Hutchins's example, this allocation of function represents a distribution of cognition through the world in which the person is acting, e.g., artefacts in the world are used to contain information or to structure the problem, other people in the world are used to share work on solving the problem or to point out information. One question that arises from this, is whether the artefacts serve to amplify human ability (see Chapter 1). Clearly it is possible for a navigator to perform the action manually (condition a) but using artefacts can change this component of the task. It would seem to me that the artefacts allow an existing action to be performed more efficiently. Thus, on one level, some degree of amplification can be seen. However, it is equally obvious that the artefacts are not merely improving the person's ability to perform an existing action, but are rather modifying the way the action is approached, conceived and performed. In this manner, the artefacts are working both physically, as external representations, and cognitively, as a means of structuring the task.

In the discussion of blacksmiths, in Chapter 5, we noted that the selection of tools for an activity and the layout of tools in a workplace represented some knowledge of the work to be done. The tools become important markers of both the type of work to be done, and also the sequencing of that work. From the perspective of cognition, even the most mundane tool provides the user with a means of defining what actions to perform and the manner in which a task should be approached. Once the tool has become integrated into a sequence of movements, i.e., once the user is adept at manipulating the tool, then attention can be directed away from motor engagement and to cognitive engagement, either in the task itself or to other, parallel activity. Consider the example of using chopsticks: the novice user will wrestle with the sticks, trying to pick up food and concentrating on the relationship between food and chopsticks. The focus of the attention will be away from her dinner partners, or will require intermittent attention to partners and attention to food. For the experienced user of chopsticks, it is possible to divide attention, such that the majority can be on one's dinner partner. The point is that some tools for some users can function as obstacles to performing the task, primarily because their use demands a great deal of attention.

Artefacts and human performance

There are artefacts that have been made to enhance human perception. For example, telescopes increase the range over which a person can see, and telephones increase the range over which a person can communicate. These devices can be thought of as 'amplifiers'. On the other hand, spectacles and hearing aids serve to provide the wearer with a level of perception that is taken for granted by other people; in other words, these do not necessarily amplify, so much as 'correct' perception. In these cases, the artefacts could be seen as amplifying failing capabilities, but are better described in terms of redefining the information that the person is presenting. There are other artefacts that do not obviously amplify, but do augment human ability. For example, a calculator augments human ability to manipulate numbers and mathematical concepts. In this respect, the calculator substitutes for a set of cognitive abilities. There are other artefacts that can represent knowledge or parts of a task, e.g., as we saw in the examples of navigation earlier.

Applying these points to the discussion of tools, we can see that some tools amplify human ability. Thus, a hammer amplifies that ability of the human hand to hit something (the same would be true of sticks, stones, clubs, axes, adzes, etc.). In this respect, the artefact is used to increase the power that can be exercised by the user. Thus, the tool represents both an extension of the person and also an amplification of ability. On the other hand, a saw (or any cutting instrument) represents a substitution, in that the tool performs tasks that the person might find difficult to do. Obviously, one could attempt to split wood using one's hands, by jumping on it, etc., but the saw provides an easier way of performing the task. The saw also provides a greater degree of control of the outcome of the activity, as does the hammer. Thus, tools provide a set of constraints on task performance, both in terms of definable outcome and also in terms of structure of performance. From the point of view of representation, Norman [4] makes the point that a shopping list is not simply a written list of things to buy. Rather, the construction of the shopping list, in terms of checking the kitchen cupboards and thinking about next week's meals, is an important aspect of the use of the list. The preparatory activities, of checking, planning, etc., result in the construction of the list, and are as important as the actual use of the list in the shops. The list could be taken to the shops by the person who wrote it, or given to someone else. Once at the shops, the list could be used in several ways, e.g., one could follow the list sequentially and find items as they appear on the list, or one could check the list as one walks down each aisle to search for relevant items, or one could use the list as set of basic requirements and add other items from the supermarket shelves, etc. Each instance of using the list might require additional activities, e.g., for the shopper who follows the list sequentially, it might be important to write the list to take account of the layout of the supermarket, e.g., by placing all fresh produce together in one column, all tinned goods in another, etc. The point of this example is simply to point out that the 'shopping list' functions as a cognitive artefact on a number of levels, e.g., as a set of requirements, as an *aide memoire*, as a plan for the activity of shopping, etc.

As Norman points out, cognitive artefacts allow activity to be distributed in several ways. First, the shopping list example shows how the activity can be divided over time, e.g., with the list being written prior to going shopping, or with the list being constructed over the course of a week (perhaps with a board in the kitchen being used to write items when they run out). Second, the shopping list divides activity across locations, e.g., the kitchen and the supermarket. Third, activities can be divided across people, e.g., one person could write the list and another could go to the shops. Fourth, activities can be divided across artefacts, e.g., the shopping list represents requirements, but the supermarket can expand this set of requirements by providing additional goods. In addition to distributing cognition, the shopping list can also structure the manner in which the task is performed. Thus, we noted different ways of writing the list and different ways of shopping using the list.

One of the ways in which cognitive artefacts play a significant role at work is in the form of procedures and checklists. The basic idea is to provide workers with a systematic guide to performing a set of tasks, and to provide a means of confirming that these tasks have been completed. Thus, maintenance and inspection activities might require workers to follow a set of tasks, in a specified order, and to sign-off each task. It is a common observation that workers do not always adhere to the procedures. For example, it might be the case that an inspection task is performed as the worker sees fit (based on a great deal of experience and knowledge of the work) and then the maintenance report is signed off in its entirety. In other words, the procedure would be to perform in a 'read task – perform task – sign for task – repeat until complete', but the actual performance would be 'do all tasks – sign the form'. If one changes the manner in which activities are supported, then task sequence can change. For instance, we developed a wearable computer that would allow paramedics to enter patient details during treatment [5]. A head-mounted display showed the values that needed to be collected, and a speech recognition system allowed the paramedic to enter the values during treatment. When the paramedics performed the tasks under 'normal' conditions (without computer support), the collection of values varied between individuals, i.e., the order in which the values were collected tended to depend on where the paramedic was kneeling (if they were near the hand, they would take the pulse), and what treatment was being administered. However, when wearing the computer, all paramedics entered all values in the same order. On the one hand, this implied that the technology led to a more systematic approach to the collection of data, and could reduce the possibility of missing values. On the other hand, this implied that the opportunistic collection of data, e.g., in terms of interleaving taken of measurements with other activities, was lost. This latter point would then have implications for how the task was performed. For example, in a study of vehicle inspection, it was found that the use of the wearable computer significantly extended the time taken to perform the task (in comparison with normal operations), and that the prime reasons for this were that the wearer had to constantly adjust the head-mounted display and that the procedures were followed rigidly in the computer condition.

Activity flow

In a study of typesetting and layout of newspapers, Bødker [6] examined the manner in which work was performed using 'conventional', i.e., non-computerized, systems. A key concept in her work was that of 'activity flow'. Some activities seem to have a fluency and cohesion that provides the performer with a sense of direct engagement in task. When the task breaks down, e.g., when there are errors, interruptions, disruptions, etc., then the flow is lost. An obvious example of this sense of activity flow can be seen in writing: when I was writing the initial paragraphs for this chapter, I was typing quickly and felt as if my thoughts were automatically appearing on the computer screen, and then the telephone rang, someone knocked on my door, I remembered that I needed to pass some documents to the Admissions Office, etc., all of which interrupted the sense of flow and meant that work on this chapter was interrupted for a few hours.

Once people reach a level of proficiency in many areas of work, there is a tendency for sequences of tasks to become automated. When driving an automobile with manual gears, it is necessary to be able to seamlessly combine the actions of declutching, moving gear shift, engaging clutch and accelerating in order to achieve a smooth change. For the novice performer, this combination of tasks can result in a whole host of problems. Thus, the integration of component tasks into a seamless sequence is one characteristic of expert performance. As we saw in Chapter 5, the expert user of tools is often able to integrate movements into smooth, seamless sequences. Furthermore, the expert is able to draw upon various cues from the environment to determine the consequences and progression of the sequence. As Barlett pointed out, '... the skilled operator did not need to examine the signals one by one, in a regular order, but could take his cues from the whole pattern...' [7].

Interrupting activity flow, therefore, raises a number of problems for the performer. However, there are many examples where such interruptions are seen as potentially useful. For example, the use of checklists to aid in inspection will interrupt the flow of inspection, by forcing the inspector to switch attention between object to inspect and paper document to read and sign.

Tools as cognitive artefacts

It has been noted that tools convey several meanings and that the tool user will be able to respond appropriately to the meanings in response to an activity. Furthermore, an activity can be planned by laying out tools and materials in a workplace, or by designing the workplace to hold the tools in specific places. Take a domestic kitchen and the activity of making a cake: you will know the whereabouts of the ingredients, scales, mixing bowl, food mixer, cake tins, spoons and spatulas, etc. Performing the activity would not typically involve laying out all of the ingredients and artefacts in one place, but could involve decomposing the activity into discrete stages. For each stage, relevant artefacts and ingredients could be selected and bought to the work area. For instance, the packets of 'dry' ingredients could be taken from the cupboard at the same time, bought to the scales and weighed in the mixing bowl, and then the packets returned to the cupboard. The next stage might involve whisking sugar, butter and eggs. The next stage might involve folding the dry ingredients into the whisked mix. For each stage, the appropriate tools and materials are selected and worked upon, and then put aside in order to move on to the next stage. Selection of tools and materials punctuates the flow of activity and helps provide a marker for the progress through a procedure. For the inexperienced cook, a recipe book might explicitly state which ingredients to collect and what tools are needed for a given stage. In this manner, the selection of tools provides the user with an indication of what action to perform and marks their point in the procedure. Thus, if the procedure was interrupted, it should be possible to resume the activity at the appropriate point. For instance, if the telephone rings when you are whisking the butter, sugar and eggs, you should be able to return to the kitchen after the telephone call and either complete the whisking or move on to folding in the dry ingredients. A fairly common source of human error would be when there is no indication of which stage in a procedure you reached prior to the interruption. For instance, you might have added baking powder to the flour and then been called away from the kitchen. When you return you cannot remember whether you have added baking powder, and proceed to add some more. One way to circumvent such a problem (and a technique popular amongst television cooks) is to place a teaspoon of baking powder on a plate, and then you know that if the plate is empty, you have already used the baking powder. In this way, the plate serves both as a holder for the measured quantity but also as a marker for a step in the procedure.

Discussion

The role of tools (and other objects) in performing acts of everyday cognition is an area that is receiving growing attention. It is clear that people develop all manner of strategies for using artefacts as external representations and markers in their activity. In some cases the artefacts have been carefully designed to support such a role. Thus, the 'three-scale nomogram', described by Hutchins, provides a simple means of performing a relatively time-consuming calculation. In other cases, the role is something that it added to an existing artefact. My proposal is that tools, whatever their design, function as external representations, and that this role can be of significant benefit in the design of workplaces. Recall the discussion of blacksmith's workshops, in which the layout of tools related to the stages of a procedure. In the design of assembly workstations (particularly in high-volume, repetitive manual work) the layout of the 'pick bins', the positioning of hand tools and the position of the operator are all designed to support a particular flow of work. This invariably makes the process more efficient (although it is hard to imagine that the work becomes any more interesting or any less likely to lead to occupational injury as a result of such a design). The point is that the layout of the workspace and the positioning of tools within that workspace provide a clear indication as to how work *ought* to be performed there. In other words, the workspace and tools provide an external representation of the appropriate procedure for that work.

10 Tools in the twenty-first century

> By its very definition the truly automatic machine needs no human assistance for its normal functioning.
>
> (James Bright, 1958)

Introduction

Over the course of the twentieth century work shifted from the primarily physical to the primarily cognitive. The expansion of computers into all aspects of work and everyday lives implies that more and more of our daily lives will involve the cognitive forms of activity that such devices require. Thus, one view of the future of tools is that almost all physical activity will become automated leaving humans with little need to physically operate upon the world around them. Such a view was a mainstay of science fiction for much of the twentieth century. One of the questions that I want to explore in this chapter is what will the twenty-first century make of tools; in other words, how will tools feature in future lives?

In order to address this question I will consider social and technological perspectives. From the social perspective, a key question relates the assumption that contemporary work is more cognitive than in previous generations. This, in turn, leads to consideration of the role of automation in deskilling of work. From a technological perspective, I want to consider current forms of supporting interaction with tools in virtual worlds, although this is not simply a matter of discussing virtual-reality applications.

Divisions of labour/allocation of function

In Chapter 5, the notion of division of labour was presented as a two-sided problem: on the one hand, there was the political question of simplifying work in order to turn the control of work from the worker to the manager, and, on the other hand, there was the technological question of allocating functions between humans and machines. Of course, there is a political dimension to the allocation of function view, i.e., as more work is allocated to the machine, so the human is further removed from the work and loses the opportunity to exercise decision-making and traditional skills.

The issue of automation and deskilling was explored by Harry Braverman [1]. His argument was that automation was simply another form of division of labour, with management seeking to maintain control of work activity through having people serve as the cogs in the machine (a view popularized by Charlie Chaplin in his film *Modern Times*: in one scene the hapless hero becomes caught in the gears of a giant machine). From an ergonomics perspective, the question of what automation does to work was nicely illustrated in a case study on the transition from traditional capstan lathes to Computer Numerically Controlled (CNC) machine operation. Table 10.1 compares these two systems.

While Table 10.1 may represent an extreme case, it does illustrate that the physical removal of the operator (in CNC) has the potential to remove the opportunity to exercise skill. However, notice that the work does not become more cognitive in the CNC condition; indeed, most of the decision-making is embedded in the software, so the operator has *less* cognitive work to do as a result of automation in this example: in other words, the operator is removed from the control loop. As Endsley and Kiris [2] demonstrate, a consequence of this phenomenon is that the operator's performance deteriorates and it becomes difficult to make effective corrections if the system becomes unstable. One point of these examples is that automation does not necessarily mean a move from physical to cognitive skills; rather, it is possible for automation to simply mean the removal of people from most parts of the work process (and a central reason for automating is to reduce labour costs). It is also worth noting that the 'traditional' operation requires a great deal of cognitive activity. As we saw in Chapter 5, just because

Table 10.1 Comparing traditional and CNC work [3]

'Skill'	Traditional	CNC
Motor	Operator sets up machine and controls during all operations, e.g., by adjusting hand controls	Operator sets up machine. Control of operations through prepared software.
Perceptual	Operator is aware of the sounds of the machine and pays attention to the cut and swarf (waste) from the object	Guard is placed over object, limiting the operator's view
Conceptual	Operator has designed the object being machined and is able to work to a template	Operator removes completed object and checks against specification
Discretionary	Operator makes decisions about type and angle of cut, selection of tool, etc., and makes moment-by-moment corrections to the original decisions	All decisions are encoded in software (with the possible exception of deciding to stop the operation in the event of problems)

someone is working with a tool in their hands, it does not mean that their activity does not have a cognitive component.

Automation, skill and organization

One obvious question to raise at this point is whether automation inevitably leads to deskilling, as Braverman suggested. In an interesting study in Germany, researchers canvassed several hundred companies that were using CNC (or similar) machinery. Part of the survey asked who held the responsibility for different functions, e.g., a programmer [P], a setter/fitter [SF] or a machine operator [MO]. Figure 10.1 indicates patterns of distribution of these functions.

Companies in which programming was performed by a dedicated programmer (i.e., organizations that used one of the first three distributions), were all large organizations concerned with mass production. Such organizations could afford to have staff working exclusively on programming. Companies with smaller workforces and with greater emphasis on flexible production tended to employ the other distributions. Thus, the manner in which technology affects work is as much a consequence of the environment in which it is used as in the design of the technology itself.

Designing for traditional skills

Whilst the organization can change the manner in which technology is used, there remains a question of how to recapture traditional skills. This was the focus of projects led by Mike Coolley in the 1980s [4]. In one project, funded by the European Union, a machining centre was designed that not only supported CNC but also allowed manual control. Figure 10.2 illustrates this concept.

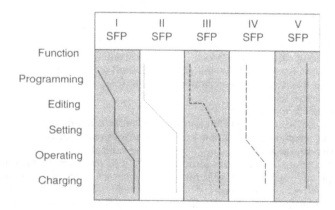

Figure 10.1 Distribution of functions across organizations [5].

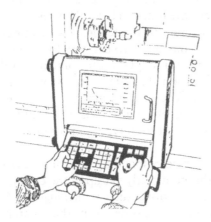

Figure 10.2 CNC+ machine [6].

In the system illustrated in Figure 10.2, not only does manual control call upon traditional lathe operation skills, but it is also possible to both write and edit software through manual operations. For instance, if the machine was running and the operator began to move the controls, the position and movement of the controls could be recorded and stored as a programme that the machine could reproduce. This is similar to the teach-by-example of several industrial robots, in which an operator would move the robot arm through the sequence of movements required to perform a task, and that arm would replicate and repeat this sequence. Alternatively, if the machine was running a programme, the operator could modify the movement of the tool by moving the manual controls which, in turn, changed the software. This example suggests that it is possible to develop technology that builds on traditional skills, rather than always imposing the need for new skills.

Virtual tools

The humble computer mouse represents a form of virtual tool, or rather it allows the user to manipulate all manner of representations of tools on the computer screen. Generally, the use of a mouse (or other pointing device) to select and manipulate objects on the computer screen is known as direct manipulation.

Direct manipulation

In a direct manipulation interface, an object on the monitor represents either a function or some form of data storage. Thus, an icon showing a piece of paper, e.g., 🗒 represents a File, and selecting this icon, e.g., by double-clicking on it, leads to the File being opened. Alternatively, an icon showing a disc, e.g., 💾 would represent the function Save. The best way to think of direct manipulation is as a reaction to the tendency of computer systems in the 1970s to employ textual commands to perform functions. It was shown that people had difficulty in recalling both the content and the structure of textual commands, which meant that computer use was often the province of the well-trained expert. The use of icons to replace text provided users with a means of recognizing which functions they could perform, and the language of interaction changed from typing-in commands to clicking on icons.

In one sense the use of icons, and the corresponding selection devices, such as mice, meant that users were able to directly act upon the computer (or at least to feel as if this were the case). Actions, such as clicking on a File icon, led to immediate (or near-immediate) responses from the computer. The fact that the icons could be selected and moved around the screen, as well as being clickable, meant that the user was able to manipulate the icons.

Like many terms in computing, direct manipulation is both a good description of what is happening and a misnomer: the user does not directly manipulate anything, rather the manipulation (of a very limited range of actions) is performed indirectly, using a device to move a cursor on a screen onto an icon.

Virtual reality

I once saw a demonstrator of a virtual-reality system that was designed to support training of assembly line operators. The demonstrator focused on the task of 'piston stuffing', in which pistons are fitted into an engine block and then tapped into place using a mallet. The user gripped a 3D mouse in one hand (in the grip one normally uses for such a mouse) and made 'hammering' motions in space. The virtual hammer (seen through the immersive, head-mounted display) then moved accordingly up and down. When the head of the virtual hammer was near a virtual piston, the piston moved into the engine block. The most obvious question that struck me, relating to the demonstrator, was 'what could the operators actually learn?' The demonstrator did not provide a 'real' interface with the tools being used, and so could only illustrate the *procedure* rather than the *practice* of doing the task. The question, therefore, is what would the 'real' interface look like to the demonstrator? I would suggest that it should look (and feel) like a hammer.

The main arena in which serious consideration is given to the sort of devices that the user might hold and use in order to interact with computers, remains the gaming world, particularly console and arcade games. Thus, if you are playing a motorcycle game, you can sit astride a mock motorcycle; in reality, you are sitting on a joystick and using your body to move the joystick from side to side in order to control the game (a small joystick, gripped between forefinger and thumb might provide the same degree of control, but offers a far less sense of physically interacting with the game). My feeling is that the games' designers are offering their customers a far richer sense of interaction than the computer world is currently considering, and that one of the reasons for this can be considered in terms of the forms of engagement proposed in this book. In the world of desktop computers, interaction devices function primarily as a means of moving objects on the screen and can be considered in terms of morphological engagement (the shape of the mouse), motor engagement (the movement of the mouse) and perceptual engagement (pairing cursor movement with mouse movement). In the world of arcade games, it is equally likely that environmental and cognitive engagement become relevant, e.g., in terms of using one's entire body to operate the control and in terms of learning strategies for effective device manipulation. For example, even with the most basic of arcade games, in which the player moves a 'ship' by spinning a trackball and 'fires' by pressing buttons, many players become so immersed in the game that they play with their whole bodies – leaning into the game when shooting, ducking and shifting their upper body when 'enemy ships' fly towards them, etc. From this point of view, one sense of immersion would be the level of environmental engagement that a person achieves during the performance of a task.

Many arcade games emphasize the notion of environmental engagement by providing all manner of auditory and physical cues to the player. To return to the motorcycle game, as the player moves through the world of the game, so rumbles, bounces and other vibrations are passed through the seat to provide a 'realistic' feeling of actually riding the motorcycle. In other words, the games are providing players with haptic interfaces.

Haptic interfaces

Buxton [7] takes the view that almost all forms of human–computer interaction contain a haptic component. Thus, moving a mouse, using a keyboard or selecting an object on a touchscreen would all contain haptic elements. Certainly one can appreciate that the user will receive kinaesthetic feedback, from moving the body part, and some degree of tactile feedback, from acting on the devices. Having said this, such feedback is seldom designed into the conventional versions of these devices. In recent years, however, there has been increasing interest in redesigning devices to provide haptic feedback.

There have been developments in haptic interaction devices, including modified mice [8]. The common aim of these various devices is to stimulate wearers in such a way as to convince them that they are interacting with a physical object. The basis for such work is that objects on a computer screen are defined with a surrounding field [9]. As the cursor enters this field, the haptic information presented to the user can be modified, e.g., actuators on the body of the mouse can be used to press against the users' fingers. These devices can be considered as 'texture mapping' devices, in that they imbue objects on the visual display with textural properties. Interacting with these devices enables users to perceive textural properties, which has been shown to improve some aspects of performance [10].

In these forms of HCI, the 'semantics' are dynamic and far more closely linked to the pragmatics of interaction. Users receive moment-by-moment feedback on the status of the object, which in turn allows them to deduce alterations in the objects' properties. Contemporary haptic interaction techniques tend to focus on subsets of the available range of haptic experiences. This means that haptic interfaces tend to convey partial representations of objects. It is not evident that the representations are necessarily the most informative or useful for people using the technology. Rather, the representations employed owe more to technological possibilities than user requirements. Virtual reality has seen the development of force-generating devices to provide programme control over haptics. One approach is to provide multiple point force input, e.g., using modified gloves which use miniature solenoids to provide multipoint vibration, piezoelectric coils, pneumatic air bladders, etc., to stimulate different areas of the hand. Another approach is to control force or position at an end-effector, such as the fingertip or stylus. By constraining motion it is possible to simulate dynamic properties of the stylus–object interface (see Figure 10.2).

In a study of suturing in virtual space, Moody *et al.* [11] used a force-feedback device (the Phantom®) to provide a feedback to the user when a suture needle punctured the skin and when the thread became taut. In order to simulate medical activity, the stylus of the device was modified to operate using a pair of needle holders (see Figure 10.3). Participants were asked to make a series of stitches in order to close the wound shown on the monitor.

The results demonstrated that force feedback reduced the time taken to complete the suturing task, increased the peak force application and improved the straightness of the stitch. The provision of force feedback

Figure 10.3 Virtual suturing.

enables the user to judge the level of force required to insert the needle hence performance becomes more efficient. Attention can then be directed to the accuracy of the task, which, in turn, means that the stitches produced were straighter. From this, we deduce that providing people with objects that 'feel' real can significantly enhance performance. However, the question of how to define the feeling of such objects is a major obstacle for virtual-reality research at the moment. An alternative approach is to interface real objects (which, by definition, possess the tactile properties with which people are familiar) to computer environments.

Real objects in virtual spaces

In the Digital Desk project [12], a person could sit at a desk with paper documents spread out in front of them and be able to mark, manipulate and read the documents as one normally does with paper. What makes the Digital Desk interesting is that a camera and projector system were mounted above the desk; thus the documents and the user's hands could be tracked to allow near-seamless integration of the world of real objects, i.e., paper documents, pens, etc., and the world of virtual objects, i.e., data and images stored on a computer. Imagine that you are reading a paper that shows a graph of performance changing over time, and you have conducted similar work – you want to compare the two graphs, so you call up the graph on your computer and have it overlaid on the graph in the paper. In this, and related applications, the presence of physical objects is registered and processed by the camera system, and then used to call virtual representations or to store images of the real objects. Thus, if the user picks up a pen and writes on the paper, the writing can be captured, processed and stored as an annotation to the electronic version of the document.

There has been growing interest in the role of physical objects in HCI. This requires development of interaction devices that support a greater range of physical movement than conventional devices, and for the objects themselves to convey information to the user. Ishii and Ullmer [13] call these physical objects 'phicons', i.e., physical icons that can be contrasted with (visual) icons used in conventional computer interfaces. One can consider phicons in terms of three categories: phicons that support interaction; phicons that display information; and phicons that store information.

Phicons that support interaction

Objects that can be picked up and manipulated, and through which the user can control the computer provide new forms of human–computer interaction. The physical attributes of real-world objects can constitute a form of display interface for computer-based tasks. In developing a 'tangible geospace', Ishii and Ullmer [13] track user interaction with real objects (such as plastic models of buildings on a campus), and map these tracked data into a computer world (e.g. to alter the perspective of a projected map display). The BUILD-IT project [14] also utilizes plastic models of buildings as a means of interacting with an architectural design package.

Phicons that display information

Objects that can change their behaviour when some critical parameter in the system changes provide novel display media. For instance, several laboratories report the using of ribbons suspended from the ceiling or pinwheels to 'display' network traffic, i.e., as a network becomes busier, so the ribbons or wheels spin more quickly.

Phicons that contain information

Objects can be used to either store data or to simulate the storage of data. Contactless data carriers, built into watches or badges, are already being used for automatic fare collection in public transport. Cooper *et al.* [15] discuss the development of an electronic wallet. They show that wallets are not merely used to carry financial items, but also represent storage devices for the necessities of everyday life, e.g., season tickets, bus passes, etc., as well as personal mementoes. A contactless data carrier can help make data transmission between devices more transparent and give these hidden operations again some link to real objects and the real world. The MediaCup [16] allows a group of individuals to coordinate impromptu meetings on the basis of shared coffee breaks. It uses a temperature sensor to detect whether the cup contains hot liquid, such as coffee, and a radio transmitter to inform other cups on the network that it is being used for coffee. The other cups can indicate, e.g., via LEDs, to their users that someone is drinking coffee and that they could go and join them.

Recent developments on the theme of augmented reality systems seek to recognize objects in the world, through the use of labelled tiles that cause the computer to project specific images [17]. For example, a tile showing the symbol ß would be recognized by the vision system, and an image of a clock face projected onto the tile. In a similar manner, Ullmer *et al.*'s [18] concept of mediaBlocks uses small, tagged blocks that are linked to digital media. Placing the mediaBlocks in a rack calls the digital media to be 'played' on a screen. The mediaBlocks are, consequently, more than merely 'control' devices; they can capture, transfer or playback the digital media, acting as storage devices as well as editors. For example, a video clip could be played when one block is in position and a second block could be used to capture this clip, then a third block (perhaps containing video controls) could be used to edit the clip. In this way, physical objects become both the repositories and means of controlling digital information.

How does using real objects affect human–computer interaction?

In the applications of table-top augmented reality, user activity is primarily concerned with moving objects on a 2D surface. Consequently, this is similar to the notion of using two mice or pucks on a surface to interact with a computer. However, there are two reasons why tangible interfaces extend the dual-mouse paradigm:

(i) it is possible that the objects themselves can offer additional cues and feedback beyond the standard shapes of mice and pucks;
(ii) the objects can offer additional functionality to the user.

In some early work into virtual reality at Birmingham, we examined the manipulation of real-world objects to support interaction with objects in virtual environments [19]. Participants use wooden discs to manipulate representations in virtual space to solve the Tower of Hanoi problem (see Figure 10.4). The study took place in a visually immersed environment, and it was suggested that the wooden blocks provided users with a range of haptic cues that facilitated their performance, e.g., participants using the wooden blocks could use the impact on block against pole to signal arrival at the destination. A further explanation for these differences in performance relates to the marked differences in strategies for object handling.

Figure 10.5 shows the movement traces to perform the task of moving one disc from one pole to another. Movement is from right to left, using the smallest disc. The 'real object' condition presents a much smoother trajectory. The 'instrumented object' (IO) represents the haptic augmentation condition and is both slower and less smooth, e.g., notice the 'bump' on the left-hand pole.

When using real objects participants would switch their visual attention to the end point of the move and not track the movement of the objects in space; whereas, in the 3D mouse conditions, participants' head movements

Figure 10.4 Virtual Tower of Hanoi with real blocks.

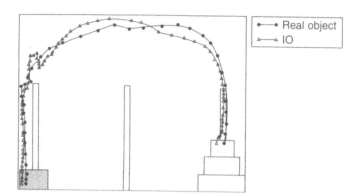

Figure 10.5 Comparison of movement traces for one participant.

followed the path of the virtual object. In other words, when using the 3D mouse the task was one of 'dragging' an object between two points on the screen, whereas when using wooden blocks the task was one of moving a real object to its new location. What this study suggests is that using real objects leads to differences in performance that are not merely adaptations of human–computer interaction techniques, such as more efficient dragging, but show use of everyday actions in the computer domain.

Discussion

One way to consider tools in the twenty-first century is as virtual objects that users can manipulate and interact with using conventional interactive devices, such as mice. Another way, is to think of future interaction devices as 'tools', i.e., physical artefacts which can be manipulated and which can convey meaning in much the same way as hand tools. In a previous book I proposed that '...interaction devices can be developed as significant components of the computer systems, not only acting as transducers to convert user action to computer response, but communicating all manner of feedback to the user and supporting a greater variety of physical activity' [20]. The point of this statement is that one can develop means of interacting with computers that extend beyond the conventional keyboard/mouse concept and which incorporate everyday objects.

From this point of view, the tangible objects in the world retain their everyday forms and meanings, but can become incorporated into computer space. By way of illustration, assume that a maintenance engineer is asked to repair a faulty valve in a petrochemical plant. The engineer's toolbox has a display that informs him of the problem and also provides instructions on how best to get to the location of the problem. On arrival, the engineer is presented with the required procedure. He takes a spanner from his toolbox, and fixes it to a nut. The removal of the spanner is logged, by the toolbox, the fixing of the spanner to the nut is logged by the spanner, as is the torque exerted to remove the nut. When the spanner is replaced, all of the data are compiled into the maintenance report that the toolbox provides to the engineer at the end of the job. Very little in this scenario is technically demanding, which means that it is perfectly feasible now. The pressing questions relate to issues of job design and other social factors, e.g., how would you react to the idea that your tools were 'spying' on you?

In this chapter, examples from CNC machines, haptic interfaces, virtual/ augmented reality have been presented to illustrate the idea that people can act on the world of computer objects in much the same way that they can act on the real world. This raises a number of possibilities, relating to both the ranges of skills that people can bring to bear in their interactions with computers and also on the ways in which computers will be situated in the world. The underlying assumption for many of the ideas in this chapter are that activity with everyday objects in the real world can lead to changes in the computer world. Thus, the operator of the CNC machine depicted in Figure 10.1 would be able to employ traditional machining skills, to use programming skills, and to combine these into a system that writes software on the basis of the user's actions.

11 Towards a theory of tool use

... from the moment of birth we are immersed in action, and can only fitfully guide it by thought.

(Whitehead, 1938)

Introduction

One of the motivations for writing this book was to develop a theory of tool use. Readers might be wondering why 'tools' should be treated differently from any other object that people use or interact with, and why 'tool use' warrants a theory. My feeling is that many disciplines are interested in ways in which people use tools, either to gain insight into human cognition, or as the basis for designing and evaluating tools. For instance, ergonomics, as a discipline, is often viewed as a body of knowledge (taken from many disparate disciplines) rather than a specific theoretical position. Having a theory of tools and tool use might supply ergonomics, and related disciplines, with a way of unifying the body of knowledge upon which it draws into a single, coherent whole.

If we return to the discussion of 'tools' in Chapter 1, we can recall that a tool is an extension of its user. The notion of extension refers to both a morphological extension, i.e., the hand holding a hammer is 'extended' in terms of its length and mass. However, extension also refers to a motor extension, i.e., the hammer allows the user to 'extend' the range of hitting action, to apply more force or to apply force in a more localized manner. Thus, one factor that differentiates tools from many other manipulable objects is the ways in which the user's abilities are extended. The notion of extending abilities calls to mind the purposeful nature of tool use, which can also be contrasted with some other uses of manipulable objects.

Cognition

The title of this book is *Cognition and Tool Use*, and cognition has featured heavily throughout the discussions. I suppose that the four key assertions relating to my presentation of cognition and tool use are:

1   tool use can be considered to have a cognitive dimension when the user has the option not to use a tool;
2   tool use can be cognitive when the user is able to choose from a set of possible tools;
3   tool use can be cognitive when the user is able to modify tool-using actions to exploit environmental constraints;
4   I do not see the fact that a tool user is able to pair a given tool with a given goal as necessarily 'cognitive'.

Each assertion warrants explanation as each is potentially counter-intuitive. Let me start with the fourth assertion, i.e., that pairing tool with goal does not require cognition. By this, I mean that one can see many aspects of tool use, particularly in fish, insects or birds, in which an object is inevitably and invariably paired with a goal. Consider how a mud wasp uses a stone to tamp its burrow; in this example, the mud wasp does not need to consider, plan or otherwise cognize either the nature of the task or the function of the 'tool', but rather works according to a consistent and instinctive set of actions.

Cognition becomes more important when the path to a goal is blocked and the potential tool user needs to frame the problem in such a way as to 'see' that a tool would be a useful and beneficial object to employ. This leads to the first assertion that cognition is apparent when the user is able to not use the tool. We saw that faced with the same objects and the same circumstances, some animals will persist in tool-using behaviours, even when this may not be appropriate, e.g., Egyptian vultures would persist in using a stone to hammer an egg-like object even if it was not an egg. It would seem that using the stone had become so much a part of the activity of opening an egg, that the mere presence of a recognizable object served as a cue to the actions.

In opting not to use a tool (or being in a position to make such a choice) the user demonstrates an ability to reframe the problem, in order to consider alternative strategies. This leads to the second assertion that cognition supports the choice between alternatives. For chimpanzees (and some crows) this might mean selecting and modifying found objects in order to perform a task. Furthermore, chimpanzees and crows demonstrate an ability to use more than one tool and to select a tool on the basis of task requirements.

It is only a small step from appreciating that different tasks require different tools to the recognition that one can manufacture specific tools to cope with the variations in task requirements, and then to the notion that one can produce whole sets of similar tools that vary in specific dimensions. Thus, a chimpanzee might select and strip twigs of specific dimensions for ant dipping. This leads to the third assertion, that cognitive tool-use involves the user being able to exploit environmental constraints. This refers both to the selection and modification of tools for specific purposes, and also to the ability to make small changes to the tool-using actions that are appropriate to the changing demands of the task. Thus, using a stone to smash a shell, e.g., by a sea otter, might involve similar levels of force and similar sequences of hammering actions. Whereas, using a stone to crack a nut, by a chimpanzee, might involve greater variation in the force of the blows, depending on the type of nut and the current state of the nut.

From this discussion, it should be clear that my concern is not to ask 'where' should cognition appear in the description of tool-using activities, but to consider how responses are organized, coordinated and represented as each of these implies some level of cognitive activity. Furthermore, the specification of a response is most likely to be learned or at least acquired through experience rather than an instinctive response. In so far as learning is a cultural activity, it is plausible to assume that the definition of acceptable (or 'proper') modes of interacting with objects will reflect cultural norms and practices.

At one level, specifying a goal is a simple matter of defining the outcome of an action. However, for tool use this may well require knowledge of the properties of the tool that one is using, the materials with which one is working and the constraints imposed by the environment. This might mean that the 'goal' involves more than simply specifying an end point for action, but involves iterating the goal as one is performing a task. For example, the goal: <divide a piece of wood into two> could be performed in several ways, e.g., by splitting it across one's knee, by leaning it against a raised surface and jumping on it, by sawing it, etc. If the goal was more specific, e.g., <divide a piece of wood into two pieces of exactly the same size>, then more control is needed in the action. It could be proposed that this level of specification, for the most part, is one way in which humans and primates differ; humans can have more specific goals. Of course, the reader might recall earlier discussions, in Chapter 3, about how chimpanzees are able to select twigs of sufficient diameter to fit into termite (or ant) mounds. One argument that some researchers have put forward, is that the chimpanzee, therefore, is able to produce a precise specification of the required twig, i.e., in terms of diameter. However, my argument is that a notion of 'affordance' was sufficient to explain this phenomenon, i.e., the chimpanzee is able to 'see' the appropriate size of the twig and selects it on the basis of perception rather than cognition. If cognition was involved then one might expect to see chimpanzees engaged in purposeful collection and grading of sticks for future activities, and this does not occur.

In conclusion, specifying a goal requires cognitive processing, but does not need to be completed entirely prior to performing the task. As we saw in Chapter 5, many skilled users of tools rely on feedback from the work in order to correct and modify their actions. In Chapter 1, everyday cognition was considered and it was proposed that many aspects of our everyday lives can be run using schema of stereotyped actions. As Norman points out, 'Everyday activities are conceptually simple. We should be able to do most things without having to think about what we're doing' [1]. It is proposed that the nature of cognition in tool use (and other aspects of everyday cognition) is such that it both allows us to perform simple actions 'automatically' and also provides us with sufficient means to monitor, check and otherwise coordinate these actions.

Environmental and morphological engagement: types of affordance

I have used the term 'affordance' throughout this book. The primary definition relates to the environmental engagement and links to direct perception. There is strong evidence, for neurological scanning studies, that the parietal cortex plays a major role in the coordination and control of object-directed action [2]. We saw, in Chapter 8, that damage to this area was related to some forms of apraxia. Consequently, we can assume that information about objects and the actions required to use these objects is represented in the parietal cortex (as well as other sites, such as the premotor cortex). This is true of both humans and monkeys, and suggests that the physical appearance of an object that can be handled is sufficient to begin preparing grasping and manipulation responses.

Consequently, one can see some support for Gibson's [3] proposal of affordance as perception–action coupling, which proposes that the appearance of an object 'suggests' a particular response, i.e., the shape of an object would evoke a particular type of grasp. Recall the discussion of 'affordance' proposed that, prior to grasping an object, the user will have decided how to grasp the object in terms of its apparent weight, balance, handle design and prospective use. The notion that neurons fire in response to the appearance of a 'usable' object implies that much of the decision is 'wired' into the neural circuitry of the brain.

Of course, when we say that an object 'affords' an action what we really mean is that, in the words of Gibson, '... perception is an invitation to act ...' [4]. For tools, this would involve two types of action: grasping the tool in order to use it, and actually using the tool. From a Gibsonian perspective, the sort of visual, auditory, tactile feedback that a craftworker obtains from the process of using tools (see Chapter 5) would be viewed as a way of modifying the world in order to perceive change; in other words, rather than the feedback being simply a means of eliciting a reflex, the very act of using the tool would be geared towards generating an expectation of feedback. Put another way, when one hits a nail with a hammer, the act of banging the hammer onto the nail's head sets up expectations about the force of contact between hammer and nail, the movement of the nail into the wood, the transmission of force and movement through the hammer to the hand, etc., and when the nail hits a knot or the hammer does hit cleanly, the expectation is violated and we 'sense' that something has gone awry.

In this book, 'affordance' has grown to mean different types of affordance. In broad terms, I am using the term 'affordance' to mean the acquisition of information at a level that need not involve conscious awareness or processing, i.e., to reflect the impressions, feelings and sensations that one might accrue through interacting with objects, people and the world around us.

Recent research has been concerned with the manner in which people acquire information about objects, e.g., through exploration of the objects. Work by Lederman and Klatzky [5] documents ways in which people manipulate objects when asked to recognize or classify them. They noted consistent patterns of finger movement and suggested that people maintain a limited repertoire of such exploratory procedures (EPs) which they tend to use in a relatively fixed sequence when faced with acquiring information from an object. This research suggests that people employ a set of EPs which are used to acquire specific sorts of information concerning specific object properties, e.g., the weight of an object is evaluated by hefting the object in the palm. One implication of this work is that we have stereotypical routines for conducting simple manipulations of objects. This is, of course, similar to the notion of population stereotypes for the control of devices, discussed in Chapter 1.

Jeannerod [6] speaks of a 'pragmatics' of human action, which is concerned with the rapid transformation of sensory information into motor command. From this perspective, the relationship between the object being manipulated and the actions that a person performs on that object become intertwined. For example, a door handle requires a specific type of grip and one can observe people orienting their wrist and shaping their hands prior to making contact with the handle. The grip used will depend on the person's interpretation of which way the handle needs to be turned. In this instance, the 'semantics' of the interaction will relate the physical appearance of the door handle to internal representations held by the person. The notion of what form these internal representations might take presents a significant challenge to the study of cognition. An obvious way of considering internal representations is to say that people can transfer manipulation techniques and activities from another domain. The reason why users would want to map experience between domains is that successful handling and manipulation of an object relies on prediction of the object's properties and of the environmental influences acting on the object.

Imagine picking up a glass of beer. In reaching for an object in front of the body, arm movements transport the hand into the neighbourhood of the object. At the same time the hand is oriented and the thumb and fingers are extended to provide clearance for the object's boundaries with the fingers following a trajectory towards surfaces that afford stable grasp. Jeannerod [7] demonstrated a degree of separability of hand transport and aperture components (although he noted that maximum hand aperture occurs at the onset of deceleration of hand transport). Thus, he observed changes in object width result in changes in hand aperture but do not affect the speed of the hand towards the object. He argued that these two aspects of movement reflect the control of separate visuomotor channels in the brain.

In addition to making the hand an appropriate shape to grasp the glass, one will exert sufficient frictional force on the glass to overcome the load forces that will act when the glass is lifted, e.g., due to the combined weight of glass and beer. With a wealth of sensory information available from tactile receptors in the fingers and thumb, you might believe that picking up the glass would involve feedback adjustment of grip force. As the arm muscles begin to develop forces to lift the hand and object off a support surface or to accelerate the hand and object from rest, the resulting forces at the finger tips might drive an increase in grip force. However, the reflex pathways involved would introduce a delay of around 80 ms and are too slow to guarantee grasp stability. Instead, in familiar situations where loading is predictable, grip force is typically adjusted in phase with changes in load force.

To study force coordination in lifting, Johansson and Westling [8] asked subjects to use a precision grip to lift objects a small distance off a support. The surface slipperiness was systematically varied (in decreasing order of slipperiness – silk, suede, sandpaper). In all cases they noted that grip force started to rise before load force. The suggestion is that people are able to rapidly modify grip forces to compensate for any changes in the stability of the object. This implies an active feed-forward interpretation of the relationship between object and grip. However, these grip forces can also be shown to be anticipatory, e.g., Turrell *et al.* [9] show that grip force adjustments in holding a tool subject to a collision anticipate the impact force which in turn reflects object mass and velocity. The former may be gained from prior knowledge, the latter from vision. If we return to the everyday experience of picking up a glass of beer (or beverage of your choice), but now assume that we are in a darkened room, say the bar of a club. We know roughly the location of our glass and can reach out to pick it up, and we know roughly how much drink was left in the glass. If the glass we grasp contains less drink and is, therefore, much lighter than we expected, then we might pick it up too quickly. This also forms the basis of a simple party trick, in which three cans (two full and a third, identical can, is emptied prior to the trick) are lined up. A person is asked to lift the two full cans and then lifts the third – on lifting the third, the person will invariably raise it too high, assuming that the weight of the can is similar to the other two. The point of these examples is to suggest that people rapidly develop and employ internal representations of the properties of the objects with which they are interacting.

The discussion so far has focused on research into object manipulation. My feeling is that this work can be worked into the argument relating to

Table 11.1 Varieties of affordance

Form of engagement	Type of affordance
Environmental	Perception–action coupling
Morphological	Exploratory procedures; hand–handle coupling
Motor	Anticipation in grip and grip-force; task-specific devices
Perceptual	Visual, tactile, auditory information from tool and tool-in-use physical characteristics, such as weight, balance, etc.
Cognitive	Supervisory attentional system; rewritable routines
Cultural	Physical appearance of tool suggests uses and, by implication, characteristics of the type of person who would use that tool.

different types of affordance. Table 11.1 demonstrates that different types of affordance can be related to the different forms of engagement.

Motor engagement: task-specific devices

An interesting perspective of the coordination of complex movement sequences was put forward in the 1990s. A 'task-specific device' outlines the machine that would comprise movement systems in order to perform a particular goal. According to the main protagonist of this idea, 'Action is an intrinsically creative business' [10]. The focus is on '...goal-directed behaviour which reveals how properties of the environment and properties of the animal are related and temporarily organized into a special-purpose machine or task-related device' [11]. This notion is appealing to the ideas presented in this book for several reasons: task-specific devices reflect goal-directed or intentional actions, they describe the relationship between environment and animal and they focus on the dynamics of this relationship. The components of such devices comprise the musculoskeletal system of the animal, e.g., muscles, tendons, ligaments, skeleton, etc., the control of this system and the perceptual system. In other words, one can think of a 'motor schema'. I realize that this is not necessarily a term that is popular with the motor skills community (particularly in terms of its apparent association with programmes or predetermined movement sequences), but it does provide a useful contrast to the notion of cognitive schema, discussed later.

If we turn this argument to a simple tool-using task, such as sawing a piece of wood. We can consider a task-specific device that would lead to optimal performance, and compare the various ways in which the device could fail with suboptimal performance by novice users of a saw. The saw will be gripped in the hand, in the form of a power grip. The wrist will be locked in order to guide the saw in a straight line. The cutting action will come from movement about the shoulder, with the elbow being pushed downwards. The person will need to stand so that the shoulder is perpendicular to the saw, in order to optimize downward force. As the saw cuts into the wood, so the action begins to acquire a rhythm, where the upthrust (or drawing back) of the saw leads to the downward thrust, as if the shoulder were being moved in a flywheel.

Bernstein [12] proposed that people adapt to repeated activity through the development of coordinative structures. A common example of this phenomenon shows the two arms of a professional tennis player; the dominant arm, that is used to hold the racket and play the shots is more muscular than the non-dominant arm. This may simply be felt to reflect imbalance of exercise of the two arms. However, Bernstein's proposal is that the dominant arm is also highly tuned to performing specific actions because, through practice and repetition of these actions, the player has been able to produce, in a sense, 'compiled routines' for these actions. Thus, the activity becomes analogous to running the routine. However, it is not simply the case that the skilled practitioner has a single version of the routine that is run in response to appropriate conditions.

In broad terms, a hard-wired musculoskeletal system, with some degrees of freedom, is 'assembled' on the basis of goals and perceptions into specific sequences of movement. When the goal recurs, the performer might seek to call upon the same sequence of movement. Having said this, it is unlikely that the performance will be identical; as Frederick Bartlett points out, in relation to playing tennis, 'When I make the stroke I do not … produce something absolutely new, and I never repeat something old' [13]. However, the repeated coupling of perception and action through practice would pre-dispose the performer to a similar task solution. When people are asked to throw objects of different weights, they will typically exhibit throwing styles that vary with object size and weight, and distance to throw. For instance, for heavier objects, the thrower might need to lock the wrist (in order to support it) which, in turn, constrains throwing style; whereas lighter objects can be held with a looser wrist and are thrown in a different style. The point is that the person can adapt throwing style to both properties of the object and perceptions of the task; when either aspect changes, so too will the action performed. As Newell observed, 'skill' is the ability to solve a motor problem quickly, rationally, economically and correctly [14].

What the skilful performer seems to do is not so much to make actions faster as to combine the actions more efficiently. In many manual skills, motor performance seems to reach an asymptote well before the performer has reached a plateau of performance. In other words, it would appear that skilled performance often arises, not from moving faster, but from other factors. These factors have been termed 'cognitive', although this is perhaps a misnomer, in that they tend to relate to the coordination of actions, e.g., in terms of the rhythm of actions, the integration of discrete tasks, the anticipation of movements, etc.

The previous discussion also highlights the possibility that sequences of movements can be repeated (although never exactly) under similar conditions. Thus, the repetition of a motor action, such as using a tool, would require the retrieval of a previously developed motor sequence. If such a sequence is absent (or cannot be accessed due to injury or damage), then the movement will have to be created 'on the fly', and might not be accurate or complete.

Perceptual engagement: interpreting feedback

The discussion to this point has suggested that manipulable objects are represented in specific terms in the brain, particularly in terms of cortical regions. So far, this would only support a 'weak' claim for specialized tool-use representations, in so much as 'tools' are manipulable objects, then they would be represented in a similar manner to, say, balls, door handles or cups. Most of the neurological studies appear to focus on static interaction with the objects, either through naming of images or through grasping of the objects. Whilst this does not reflect the complexity of tool use, it does imply that the grasp action is evoked by the appearance of the object. This, in itself, is of interest because it implies that the design of the object (in terms of its physical appearance) not only 'suggests' how to use the object but, more fundamentally, sets up neural activity that associates visual features with action in terms of an expected or anticipated response.

In order to work with an 'anticipated' response, it is necessary to be able to both represent the sort of response that one might expect and also to be able to perceive information from the act of using a tool that can provide evidence for the response. To this end, perception becomes a process of monitoring the behaviour of the tool and the material upon which it is working, e.g., through visual, auditory, olfactory and tactile feedback. It is also highly likely that use of tools involves kinaesthetic feedback that relates to the movement and position of one's body in space. However, it is apparent that a great deal of correction to movement, particularly amongst skilled practitioners, is occurring at speeds that do not sit easily with the requirement to process feedback. Consequently, it is possible that such actions involves the performance of a task with some constraints defined in the motor schema to determine when the activity is beginning to break-down. In this manner, one can correct the slip of a blade quickly and efficiently. In this respect, the feel of the tool becomes a significant contributor to the monitoring of performance. Of course, if tactile feedback is removed from the tool user, e.g., if they are wearing thick gloves or working in extreme conditions or have suffered injury that prevents perception of feedback, then performance becomes slower and reliant on visual feedback.

Cognitive engagement: cognitive schema

The notion of task-specific devices assumes that people can make use of schema. The notion of schema has had a rich history within cognitive psychology, and can be traced at least as far back as the work of Frederick Bartlett [15]. In Bartlett's view, a schema was '... an active organization of past reactions, or of past experiences, which must always be supposed to be operating in any well-adapted organic response' [16]. The point that he makes is that any behaviour that appears to be well organized will probably draw upon collections of similar behaviours. These collections of behaviours, therefore, do not exist as ordered series of actions so much as '... masses of organized knowledge' [17]. For Bartlett, schema were, by definition, not open to conscious awareness and consisted of active knowledge structures. This means that schema are not stored as static representations of series of behaviour, but are reconstructed through combining past experience with current stimuli. An appealing aspect of this notion of schema is that it implies that behaviours are built anew when a task is performed, but built by assembling components and 'chunks' of previously learned and practiced behaviours. This bears a striking similarity to some of the notions of task-specific devices considered earlier.

Supervisory attentional system

Some 50 years after Bartlett was defining his view of schema, Norman and Shallice [18] proposed a view of everyday activities that was essentially schema driven. In their model, there are two control structures that moderate and coordinate behaviour. A horizontal structure (the Supervisory Attentional System) comprises of autonomous schema that define sequences of behaviour. For instance, consider the various pieces of knowledge that go into the simple task of changing from first to second gear in a manual gearshift, e.g., ease off accelerator, depress clutch, shift gear lever, engage clutch, accelerate. For the first-time driver, this sequence can be particularly daunting (especially as it is to be performed in conjunction with a host of other tasks, such as steering the car, watching the road, checking the dials on the dashboard). With practice, the sequence is compiled or chunked into a single unit or schema and can be performed 'automatically', i.e., a habitual action that progresses without the need for conscious attention. From this point of view, one might assume that the vast majority of tool-using activities would illustrate automatic routines that require little, if any, attentional control. With practice and experience we learn to combine co-occurring tasks into higher-order schema. One way of thinking about this is in terms of a hierarchical structure, in which goals can be decomposed into subgoals which then relate to tasks, e.g., in much the same way that ergonomics uses hierarchical task analysis to describe goal-based activity. In this manner, one can 'call' a routine of subgoals when faced with a situation that has features in common with familiar goals.

In addition to a horizontal thread, a vertical thread (the Contention Scheduling System) influences behaviour by moderating such factors as attention or stress. One role of these vertical structures is to inhibit or turn off the habitual actions in order to support the performance of a novel action. For instance, consider driving in a country in which the rule is to drive on the opposite side of the road to which you are used to driving. The role of the vertical structure is to not only ensure that you stay on the correct side of the road, but also to make sure that you do not inadvertently drift to the other, e.g., when driving on a roundabout or after turning off a main road onto a side road.

One of the main implications of Norman and Shallice's notion is that the majority of everyday behaviour proceeds as planned, with little or no conscious involvement on the part of the performer. Thus, opening a door is an activity that is so taken-for-granted, that it is trivial. However, its triviality belies the necessary coordination and control of the myriad subtasks and subdecisions that are required. Furthermore, if one considers that manner in which everyday actions 'breakdown', as discussed in Chapter 8, one can see that the 'action as performed' bears some similarity to the 'action as planned' or to other plausible actions in the same context. To take a simple example from my own experience, some friends bought a cake and I made them a pot of tea to have with the cake. Having delivered a tray with teapot, teacups, sugar bowl, milk jug and teaspoons, I returned to the kitchen to collect a knife to cut the cake. When I opened the cutlery drawer, I reached in and collected another set of teaspoons rather than the knife (which was, incidentally, not in the drawer but in the knife block on the kitchen counter). The 'breakdown' in this action can be attributed, I think, to my failure to deactivate the 'make tea' plan and to not fully develop the 'collect knife' plan. The suggestion is that the planning and coordination of these simple actions is, in some sense, hierarchical. This notion of a hierarchy of control can be seen in the discussion of cognitive skills in Chapter 8 and also in the preceding discussion of task-specific devices.

In an extension of the Norman and Shallice work, Cooper and Shallice [19] propose that everyday actions can be considered in the terms of four layers of a hierarchical schema network:

1   Layer one: goals
2   Layer two: schema
3   Layer three: subgoals
4   Layer four: primitive actions

In this model, activation can occur through top-down activation, e.g., from goal to action, or from bottom-up activation, e.g., an object might 'afford' an action or, indeed, a goal, and through lateral inhibition, e.g., competing schema will need to be inhibited in order to allow the 'correct' schema to operate. When a goal is completed, the schema is inhibited – this is similar to the notion of 'popping' a goal stack in the GOMS model. In addition to a schema network, it is proposed that an object network contains information related to the representation of objects. For instance, an action such as 'pick up' might activate objects such as cup, spoon or teapot. Lateral inhibition would then ensure that competing objects are ignored. From studies into apraxia and other disorders of movement (see Chapter 8), it is apparent that some of the disorders relate to failures of representation at some level of these hierarchies, e.g., people might have difficulty in linking an object with a goal, and others to failure of lateral inhibition, e.g., people might find an object triggers an action, even if this is not relevant to the task.

Rewritable routines

In work into the design and development of technology for everyday use, Baber and Stanton [20] propose the notion of 'rewritable routines', which can be viewed as an attempt to combine some of the work on schema and stereotypes into an explanatory framework. The physical appearance of a tool (or any product that a person can use) presents information to the user in a variety of formats. The physical appearance of an artefact is called the 'system image'. Some information is permanently accessible, e.g., the visual appearance of handle, head, etc., while other information requires user action to be accessed, i.e., the dynamic properties of the tool in use. It is proposed that the 'system image' *implies* a set of routines that the person can use, and that the selected routine will depend on the user's goal and on the user's previous experience. The selection of a routine will be influenced by prior experience and by the system image. The 'system image' provides cues to the sequence of tasks. The physical appearance of the tool 'affords' a particular set of grasps and a possible set of user goals. Once grasped, the tool 'affords' a particular set of postures and a possible set of actions.

It is assumed that all interaction with tools will be goal based and purposeful, i.e., that people have a reason for using the tool and they seek to match their goals with the tool's functions. One can further assume that interaction between user and tool proceeds through a series of states. At each state, the user interrogates the system image for a correspondence between goal, form and function. Ideally, the goal, form and function would match and the action would be obvious. In the ergonomics literature, such matching is known as 'S–R compatibility', e.g., turning a control knob clockwise causes a pointer to move to the right on a linear scale (see Chapter 1). People appear to have well-developed stereotypes for some forms of S–R compatibility and to base their actions on these stereotypes. Thus, at one level of behaviour users can simply match the system image with a stereotyped response. Such stereotyped responses can be thought of as Global Prototypical Routines. Errors can arise when users mistakenly match an object in the system image with their goal, or when a strong stereotype overrides the correct action.

One rarely achieves the overall goal with a single action, and users need to keep track of their position in a sequence of goals. Further, there may be occasions when there does not appear to be a clear match between goal, form and function. At this stage, users need to infer the appropriate action on the basis of the system image. Rather than carrying a representation of the product throughout the interaction, users only require information when they are unsure of the appropriate action. The 'routine' represents a set of actions (or a single action) that is deemed appropriate for a given state. These can be thought of as State-Specific Routines. Interpretation of the system image in terms of the current goal state might draw on knowledge related to other tools, i.e., through analogy or metaphor, in order to infer an appropriate action. Once the action has been performed, then the knowledge is no longer required. In this way, the routines are 'rewritable' in that they can be overwritten by subsequent information. Thus, users might invoke one or two pieces of information from long-term memory (using metaphor or analogy) in order to determine an appropriate action. However, in order to minimize working-memory load, the users will rarely need to maintain this information throughout the interaction, and will concentrate on monitoring their progress towards the goal.

It is assumed that the majority of routines that one uses will be state specific. This is because movement through states in human–machine interaction is punctuated by brief periods in which the machine responds to user actions. From this perspective, users will employ multiple routines to determine relevant action as and when required.

Cultural engagement: representing activity

In Chapter 5, Activity Theory was introduced and used as the basis for discussing cultural engagement. A fundamental concern for Activity Theory, following the lead of Vygotsky, is the influence of culture on the acquisition and development of skills and knowledge. In their discussion of blacksmiths at work Keller and Keller [21] note that it is possible for a skilled practitioner to walk into a fellow blacksmith's shop and know what sort of work they had done, not by the finished products but by the layout of the workshop. In a similar vein, skilled practitioners can often judge the level of competence of an apprentice or novice worker simply by the way in which a particular tool is grasped.

This illustrates a particular set of values and rules governing specific branches of activity. The selection of tools, their arrangement in a workplace and the manner in which the tools are held and used, are all products of the culture of a particular domain of activity.

From a different perspective, it is possible to consider a concept of 'cultural affordance' [22]. In order to illustrate this concept, imagine a tree stump in the middle of a forest. The tree stump is the result of logging in the area, and the top of the stump (at about knee height) is flat. Now imagine a tea cup placed on this tree stump – the tree stump now 'becomes' a table. The change in role (from tree stump to table) arises from the purposive human activity to which the object is put. According to the notion of 'cultural affordance' the change in role is a consequence of the tendency of humans to bestow meaning and value upon objects in their world. In an extreme case, the meaning will depend on the knowledge of a particular cultural grouping. Thus, a collection of medieval farm implements might have clear 'meanings' for specific activities to the peasants who used them, but to a modern day observer might 'mean' either 'old farming tools' or (equally possible in the tendency to provide antique ornaments in public houses) might mean 'unusual decoration'. Modifying the various components of the 'context of use' of these objects, therefore, changes the relevant meanings that can be attached to them. The key point is that the meaning of the object is an interaction of the knowledge held by the culture that produced the object and the knowledge held by the culture that is viewing the object.

Discussion

One way in which we can gain an insight into the processes involved in using tools is to consider how experts differ from non-experts. It is well known, from the massive literature on problem solving, that experts tend to use more abstract representations of a problem than non-experts, who work at a more surface level. The experts perfect strategies for chunking related items into coherent wholes, which means that they can remember complex patterns, such as the layout of a chess board. Experts are also adept at selecting and applying appropriate 'rules' for handling problems. However, most studies of problem solving also suggest that 'expertise' is domain specific, when the problem is shifted outside of the normal domain of the expert, e.g., if chess pieces are randomly placed on a board or if the familiar rules no longer apply, then performance tends to mirror that of the non-expert.

Taking Bartlett's schema theory, one can assume that experts are simply able to draw upon richer schema, and can develop these schema more quickly and apply them more flexibly than the non-experts. Consider the trivial task of buttering a piece of bread, using butter that has come straight from the refrigerator. Spreading the butter might lead to the slice of bread being ripped and torn. One might follow several strategies to either make the butter softer, e.g., using a hot knife, spreading the butter onto the plate first to soften it, heating the butter. Each strategy might lead to varying degrees of success, but the point at issue is that, in order to achieve the goal ('spread butter without damaging bread'), it is necessary to adequately frame the task as a problem, i.e., to realize that hard butter *could* damage the bread. While I can frame the problem, my eleven-year-old daughter, for instance, might not be able to and will attempt to spread the butter despite the damage to the bread. However, she does possess the motor skills that enable her to spread butter, whereas my four-year-old son is only beginning to work on the manipulation of a knife as a cutting device and not as a spreading device (and my three-year-old twins are still at the spoon and fork stage). The butter spreading problem, trivial though it is, crosses the forms of engagement considered in this book and provides a starting point for considering the notion of expert tool-using. Furthermore, the example illustrates the two types of schema that have been explored in this chapter. A 'motor schema' governs the manipulation of the knife and the spreading of the butter. It is sufficiently well defined to allow repeatable action, but sufficiently flexible to allow moment-by-moment correction and modification to cope with the fluctuating demands of changes in situation. A 'cognitive schema' governs the planning and control of the action. It is particularly well suited to proposing alternative courses of action and to reconciling a course of action to specific situational demands.

12 Conclusions

Habit diminishes the conscious attention with which our acts are performed.
(James, 1890)

Introduction

My focus in this book has been on the different ways in which the use of tools can be considered. The principle aim has been to develop what might be thought of as a sketch for a theory of tool use. In particular, I have been interested in developing an insight into aspects of human behaviour that are typically performed with little or no conscious attention. Tool use, therefore, represents both skilled performance and everyday activity. One of the reasons why I have focused on such activities is that they are prone to disappear from our awareness and become taken for granted. Indeed, is it not a characteristic of a well-designed tool that one loses sense of the tool and becomes only aware of achieving a goal? To put this another way, we are only aware of poorly designed tools. This suggests that the proper measure for the design of a tool is the fit with the person who will be using it. Consequently, there is a role for the various sciences of human behaviour to advise and influence the design of tools. As we saw in Chapter 6, ergonomics has long provided a foundation for tool design, particularly in terms of the physical dimensions and shape of the tools. Thus, the form of tools can be related to the fit between the tool and the physical characteristics of the potential user. However, my argument is that there exists a fit between the tool and the cognitive characteristics of the potential user. This is obvious when one considers 'tools' that work upon information, such as computers, but is also, I feel, highly relevant to the study of more conventional tools. In order to explore the cognitive aspects of tool use I have developed a fairly simple notion of forms of engagement.

Forms of engagement

The selection of six forms of engagement is, of course, arbitrary. I could have opted for two (physical vs cognitive) as easily as a dozen or more.

I could have decomposed each 'form' into myriad 'subforms' or proposed any number of alternative combinations. Like any framework, my proposal of six forms of engagement provides a convenient starting point for considering a particular phenomenon. My feeling is that six forms is sufficient for the purposes of discussion in this book; I have been able to accommodate almost all of the main points that have arisen from both the review of available literature and the discussion of tool-using activities within the scope of the forms of environmental, morphological, motor, perceptual, cognitive and cultural engagement. If I have given undue weight to some forms over others this is partly a consequence of the available literature and partly a result of my own thinking about how I use tools. Given my feeling that six forms of engagement are sufficient to consider tool use, the next question is whether they are necessary. After all, one might feel that once you know how to hold a hammer to bang in a nail, you have all the knowledge you need for this task. Decomposing activity into constituent parts and producing frameworks, taxonomies and other means of classification is, of course, an academic imperative; it is what academics do and reflects, in part, their desire to put order into the world. For the study of tool use, the idea of six forms of engagement provides me with a vehicle for contrasting different uses of tools, both across species and also across types of work.

Contrasting animal with human tool-use

Table 12.1 presents a comparison of tool use across the species reviewed in Chapters 2 and 3 of this book, and against aspects of human tool-use explored through the other chapters. My intention is to contrast the species in terms of how specific forms of engagement might be applied to them. In part, I was interested in understanding why human tool-use might be different from that of other creatures. More importantly, I wanted to see whether the tentative notion of 'forms of engagement' was sufficient to allow differentiation of tool-using activities. This has also allowed some consideration of the manner in which different species are able to separate objects in the environment from the environment itself. By this I do not mean a simple physical separation; after all, it is an easy matter to pick something up and hold it. Rather, I mean a conceptual separation, i.e., to recognize that object and environment are independent and that the object can be used to affect changes in the environment (or other objects).

Environmental engagement

Let us begin with environmental engagement. From the review of tool use by insects, fish and crustaceans, it strikes me that these creatures do not need to separate an object from its environment in any sense other than to 'see' that the two things can be moved independently of each other, i.e., a stone can be picked up and moved. This does not require an appreciation

Table 12.1 Contrasting animals and humans in terms of forms of engagement

	Insects, crustaceans and fish	Birds	Mammals and apes	Chimpanzees	Early hominids	Humans
Environmental	Combined object with environment	Separate object from environment	Separate object from environment	Shape object	Fashion object	Manufacture object
Morphological Motor	Jaws Extant	Beak Extant	Hands/teeth Extant/Modified/Tool-focused	Hands/teeth Modified/Tool-focused	Hands Tool-focused	Hands Tool-focused
Perceptual Cognitive	Action Action	TOTE Act to goal?	TOTE Act to goal?	TOTE Act to goal	TOTE? Plan, monitor?	Feedback Plan, monitor
Cultural	—	—	—	Group specific	Shape of stones?	Rules, praxis, community

of how the separation can be affected (beyond the ability to raise the stone). On the other hand, birds and apes are clearly able to act upon objects in order to separate them from their environment, e.g., removing twigs from branches. The separation of object from environment is performed in order to reach a goal. For the insects, crustaceans and fish it appears that the manipulation of the object *is* the goal; in other words, that the intention to perform an action is motivated as much by the presence of the object in the environment as by any 'goal' that the creature has developed. Of course, birds can also be observed to display intentionality through their action, i.e., the presence of something that looks like an egg might be sufficient for the Egyptian vulture to engage in stone hammering. For humans, generally, it is only when actions are performed under unusual circumstances (i.e., under severe stress or when the person has neurological impairment) that an object would be sufficient to 'trigger' an action. This implies that some mediation must occur between seeing the object and defining an action. For chimpanzees, the separation of object from environment is taken further, in that they display ways of modifying the object, e.g., stripping leaves from twigs, and also alternative strategies for selecting the objects relative to the goal, e.g., selecting twigs of certain dimensions. This implies not only an ability to associate an object with a goal, but also an ability to interpret the value of an object. As I have mentioned in several chapters, I am not sure that this ability necessarily implies 'cognition', but that it might also be explained in terms of 'affordance' (see later). For early hominids, objects were not only modified but also fashioned for specific purposes. This implies the ability to not only interpret the value of an object but also to fashion the object so that it better suits this interpretation. For modern humans, tools are not merely fashioned, but manufactured. By this I mean that, even when producing a tool by ourselves, humans will seek to develop constituent parts, such as head, handle, etc., and then seek to attach these parts together. The idea of a tool consisting of several parts is uniquely human.

Morphological and motor engagement

In terms of morphological and motor engagement, insects, crustaceans and birds have generally recourse to only one means of engaging with the tool, i.e., beak, mandibles or claws. The tool is held in a fixed position, so that it becomes 'part' of the creature's head. In this way, the use of the tool will simply mirror extant behaviours, e.g., pecking with the beak becomes 'hammering with a stone'. Mammals and apes are able to use paws, claws and hands to wield tools. This means that the tool-using behaviours could be modifications of extant behaviours, e.g., the elephant holds a stick in its trunk and then flicks its trunk, or could be novel tool-related behaviours, e.g., the sea otter grasps a stone and then performs a 'hammering' action which is not typical activity when not grasping a stone. For chimpanzees, the tool-using behaviours, particularly when considering acts of display and aggression, could be modifications of existing behaviours, e.g., brandishing a stick could be analogous to shaking a tree. However, many of the behaviours appear to be tool-focused, albeit with some modification. Thus, while hammering with the fist is an extant behaviour, holding a stone and using this to hammer suggests both modification of 'hammering' and also the development of a new technique. The nut-cracking activity of chimpanzees also indicates bimanual cooperation to a level that is not really seen in other species (of course, birds might use beak and claws when eating or when removing sticks but tend not to show the use of beak and claw when using tools). Bimanual cooperation was also demonstrated by some of the stone-tool making techniques of early hominids and is clearly a characteristic of human tool-use.

Using two hands

So far, we have tended to focus on tool use as an activity involving one hand. However, there are many instances of everyday tool use in which the user is required to use both hands. An obvious example is the task of hammering in a nail: one hand holds the nail in place until it can stand unsupported and the other hand wields the hammer. A measure of coordination relates to the synchronicity of performance in terms of the capacity of the two hands to work together to produce a shared action and goal. Therefore it is argued that superior performance would be indicated by the ability to combine the actions of both hands into a single activity. The Kinematic Chain Model [1] describes bimanual cooperative action. The model hypothesizes that the left and right hand form a functionally kinematic chain. This results in three general principles:

1   Right to left preference – the right hand performs actions relative to the frame of reference set by the left (in right handers);
2   Asymmetric scales – the right and left hand are involved in asymmetric temporal scales of motion i.e., the left-hand actions are low frequency compared to those of the right;
3   Left-hand precedence – the left-hand actions precede those actions of the right.

We have applied the kinematic chain model to tasks involved in suturing [2]. Figure 12.1 shows data produced by three people performing a suturing task, in the form of process charts. Left-hand activity was of a lower frequency than that of the right hand. The right hand performed more skilled actions while the left hand performed gross leading actions. Skilled participants produced regular coordinated actions, with strong bimanual support from the first suture. The less experienced participants took longer to adapt to the equipment, and showed clear improvements in their technique

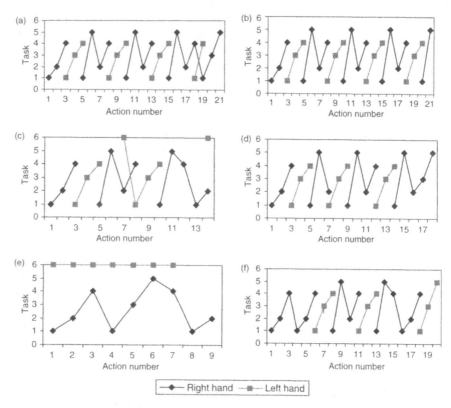

Figure 12.1 Bimanual coordination for suturing across skill level and trial.
(a) Senior registrar – first suture; (b) Senior registrar – last suture;
(c) Junior registrar – first suture; (d) Junior registrar – last suture;
(e) Nurse – first suture; (f) Nurse – last suture.

although the level of hand coordination did not reach the same standards
as the skilled participants.

Perceptual and cognitive engagement

It is proposed that perceptual engagement functions in three main ways:

1   it is either entirely embedded in the action itself, i.e., it becomes a form
of environmental engagement;
2   it is used in the manner of TOTE (test-operate-test-exit) discussed in
Chapter 2;
3   it is used as a means of interpreting and drawing upon feedback.

For the most part, insect, crustacean and fish tool-use seems to exhibit
purely environmental engagement. In other words, the mud wasp will use

a stone to tamp mud into a burrow and will tend to repeat the hammering action a few times until the mud is flat. There is no sense in which the tamping is performed, the action halted and the results checked, with a view to repeating the hammering. In other words, the tool use by these creatures tend to rely on the 'unpacking' of the complete motor programme. For mammals, birds and apes, on the other hand, tool use is characteristically performed along TOTE lines. In other words, a burst of activity will be followed by a checking of the results of the activity and then followed by another burst of activity if required. What is not apparent from this type of tool use is whether there is any moment-by-moment modification and correction of the tool-using behaviours. Rather it would appear that the bursts tend to follow similar patterns, with little variation in activity. The stone tools produced by early hominids might also have relied on a TOTE approach to control activity. However, chimpanzees and humans demonstrate a clear aptitude for both monitoring their performance while they are working, i.e., they do not need to stop working in order to test their actions, and also the ability to continually adjust the manner in which they are working.

Both TOTE and 'feedback monitoring' rely on the user of a tool having some sense of working towards a goal, which suggests intentionality. As mentioned in Chapter 2, it seems that the intentionality of activity for insects, crustaceans, fish and some birds and mammals is evidenced by the action itself, rather than by any premeditation, preparation or cognition. For chimpanzees and humans, intentionality manifests itself in the ability to choose an appropriate course of action and then to be able to modify and correct that action as required. In terms of cognition, this implies two dimensions: (i) a set of processes that can store, retrieve and develop courses of action that are appropriate to the interpretation of a given situation, and (ii) a set of processes that can monitor, edit and otherwise correct these courses of action during performance. In broad terms, I have referred to these processes as cognitive and motor schema. The errors that chimpanzees make and their comparatively limited performance suggest that the store of courses of action might be somewhat smaller than that of humans. I am not going to debate why this might be the case, although throughout the book it has been suggested that possible explanations might cover differences in morphology (leading to a different repertoire of physical skills), or differences in enculturation (leading to both different sets of beliefs about the nature of objects in the world and different explanations as to why tasks are performed in the manner that they are).

Cultural engagement

There is little, if any, evidence to suggest that insects, crustaceans, fish, birds, mammals and apes explicitly engage in the cultural transmission of tool-using abilities. Of course, this is not to deny that certain species in certain locations are more likely to demonstrate tool use than their counterparts in other locations. However, the lack of obvious enculturation in species except for chimpanzees and humans, implies that the vast majority of animal tool-use is dependent upon innate behaviour or imitation.

Developing a theory of tool use

In Chapter 11, the notions of cognitive and motor schema were considered. These provide a means of considering the manner in which tools are used. Much of the literature on either cognition or motor control has relatively little to say about tool use (or indeed, about everyday activities in general) and so the following will be entirely speculative. It strikes me that a significant problem for ergonomics arises from a distinction between the physical and the cognitive aspects of human performance (a similar issue appears to be at work in the fields concerned with the study of motor skills, with a periodic war being fought between those who believe that cognition has no role to play in the control of motor skills and those who do). I have always felt that dividing the physical from the cognitive was both arbitrary and, ultimately, fruitless. The challenge, of course, is how to bridge the chasm.

In the discussion of the breakdown of tool use (in Chapter 8), in particular the discussions of apraxia, it was apparent that 'tools' (both as manipulable objects and as objects that are used in sequences of action) have specific neurological representations. Recall that an image of a 'tool' evoked activity in that part of the brain largely concerned with motor control and that damage to the parietal cortex or frontal lobe could lead to problems in sequencing everyday actions. Further, the physical presence of a tool could be enough to 'trigger' an associated response, even when this response was not appropriate or asked for. From this it can be concluded that 'tools' are represented both in terms of their physical appearance, and the manner in which they are used. One could almost suggest that there is a representation for form and a separate one for function.

Before progressing, it is worth asking the question what makes tools any different from the countless objects that we encounter and use in our everyday lives? Surely we encounter many 'manipulable objects' on a day-to-day basis that are not tools in the sense used in this book. Obvious examples would include keys, door handles, gear-sticks, etc. My feeling is that tools occupy a space that is both related to other manipulable objects and quite distinct from them. A tool is designed and used with the purpose of acting upon the world in order to effect a change. Of course, the other objects are also covered by this definition. However, tools provide an opportunity for flexible manipulation by the user in order to control and refine the effects of the changes. In other words, tools are instruments through which fine motor skills can be expressed and which are designed to allow the user to modify and alter the manner in which the tool can be used. Thus, a door handle supports pretty much one type of grasp (with minor variation) if it is to be used properly, but a screwdriver supports at least two types of grasps, e.g., from the precision grasp to place the tip into a screw's head to the power grasp to drive the screw home. A gear stick needs to be operated with a specific degree of force if it is to allow the gears to change, but a hammer can be operated with varying degrees of force, e.g., from the gentle taps to get a nail to bite to the firmer hits to drive the nail in. The other manipulable objects considered earlier all function as specific devices to perform single, specific operations. They are, thus, defined entirely by the immediate context and by the design of their surroundings. Tools function as the means by which sequences of actions can be combined together and performed in a variety of contexts.

From this discussion, my proposal is that tools warrant special treatment. In terms of motor schema, the discussion in Chapter 11 leads to the following proposals:

1   We hold schema that define appropriate grasps for manipulable objects. The selection of an appropriate grasp typically occurs as we are approaching the object, i.e., it is not a pre-planned operation but is defined *ad hoc*. The selection will depend on the object that is to be grasped, e.g., the width of a handle, the material from which the handle is made, etc. The selection will also depend on the use to which the object will be put, i.e., raising a full glass to drink or an empty glass to put into the washing-up bowl. The definition of an appropriate grasp is, in all likelihood, part of a learned repertoire of object manipulations that we have acquired during our lives and may well exhibit similar stereotypical tendencies to the 'population stereotypes' that influence the use of controls (see Chapter 1) or the exploratory procedures that influence our testing of materials (see Chapter 11);

2   We hold schema that define appropriate coordination of actions. These have been considered, in Chapter 11, in terms of 'task-specific devices' and 'coordinative structures'. The idea is that a rehearsed set of actions begin to reinforce certain grouping of muscle firing and other neurological events. Thus, when faced with a similar situation, one will be able to cue the set and perform it rapidly and efficiently. Of course, there is ample evidence to suggest that people do not simply 'run' a programme, but that the set is open to adjustment, correction and other modifications during its performance;

3   We hold schema that define the appropriate sequencing of actions. Thus, the task of 'making a cup of instant coffee' has a clearly defined group of actions and a sequence of performance. Again the sequence need not be rigidly defined, e.g., does it matter if the step 'get teaspoon' is performed before the step 'get cup'? Having said this, some steps are clearly dependent, e.g., the step 'put a teaspoon of coffee granules in cup' clearly requires that the teaspoon, cup and jar of coffee granules are to hand.

For each schema, we know what happens if some aspect fails, both in terms of neurological damage or in terms of slips. Thus, failing to select an appropriate grasp on approaching an object could lead one to collide with the object, knock the object over or otherwise not grasp the object appropriately. If the object in question is the handle of a power tool, say, then the failure might also arise from grasping too low on the handle (so that the weight of the tool pulls the wrist forward). Having grasped the object, selection of an appropriate set of actions might fail due to incomplete specification of force, movement, etc. If the object is a tack hammer, one might apply too much force and bend the tack when hitting it. Having selected an appropriate set of actions to use the tool, then the process could fail when a step is added or omitted from a sequence of actions. If the procedure relates to a maintenance job in which a collar is tightened and then a retaining screw fitted and screwed in, it is plausible that the 'tighten collar' would serve to signify closure and the 'fit screw' be omitted.

In terms of cognitive schema, the following proposals can be put forward:

1   We hold schema that relate the appearance of objects in the world to specific goals. From this it could be proposed that one holds a hierarchy of 'object appearances' (forms), with different aspects of the appearance linking to functions, and a mechanism through which the form-function representations can be linked to specific goals. Thus, the goal of 'fix a picture hook on the wall' could be related to goals of 'fixing objects' using nails, screws or other media. The selection of path for solving the goal might well depend on prior experience, i.e., might relate to the schema that describes solving a related goal in the past. Thus, to paraphrase an old proverb, 'to a man with a hammer, all problems are nails';

2   We hold schema that relate the function of tools to previous experiences of similar tools. Thus, until we know otherwise, we might assume that all hammers are used in the same manner, or that a power screwdriver is used in a similar manner to a 'manual' screwdriver;

3   We hold schema that are shaped by our enculturation and experiences of the specific tool-set of the country in which we live. Thus, using knives and forks to eat are 'natural' in some countries, chopsticks are equally natural in others and using breads to eat food is equally natural in other countries. When we move between cultures, then either the schema require modification or we need to develop new schema.

The notion of SAS (discussed in Chapter 11) provides a means of explaining how our actions can fail, when considering cognitive schema. In general, the demands of a situation will be assessed and a relevant schema selected. If the wrong schema is selected, then the action might fail. For instance, we might decide to achieve the goal 'hang a picture on the wall' using a hammer and a nail, only to find that the bricks in the wall are so hard that the nail bends. Having selected an appropriate approach to reaching our goal, we might then fail to select an appropriate way of using the selected tool. For instance, we might decide to use a screw instead of a nail and, having drilled the hole and fitted a rawl plug, we attempt to use an electric screwdriver by turning it on the screw rather than letting the motor do the work. In this instance, we need to effectively 'inhibit' the schema that describes screwdriving in order to 'activate' the schema for using power tools.

The last example, of using a power screwdriver, might seem to be an example of motor schema. I feel that it is slightly different, in that cognitive schema tend to manage actions whereas motor schema tend to coordinate them (although I accept that the distinction is not particularly clear). A more important point is that the cognitive and motor schema overlap, and that one can see the level of control required might vary between them. This is similar to Rasmussen's [3] proposal that human performance can be skill based, i.e., well learned and automatic; rule based, i.e., requiring the application of heuristics and procedures; or knowledge based, i.e., requiring problem solving, interpretation and decision-making. The point to note is that tool use requires the application of motor and cognitive schema, and that these schema are generally applied in an automatic manner that requires little attentional control from the experienced user. As experience decreases, so the requirement for attention increases, and the manner in which the actions are performed becomes significantly altered, e.g., the occurrence of 'slips' greatly diminishes, and different sorts of errors come to the fore.

Relating schema to forms of engagement

Throughout the book there has been an implicit assumption that tool use is schema-driven. This is one of the reasons why the tool use of insects, fish, mammals and birds was felt to differ from that of chimpanzees and humans; the 'schema' used by the latter were more detailed than those used by the other species.

In Chapter 11, engagement was considered in terms of affordance, and several sorts of affordance were proposed. It seems to me that, in order to develop an appropriate response to the affordance of things an organism needs both the capability of responding to the thing, i.e., through some form of neurological hard-wiring, and the capability of organizing a response. We know that 'tools' (and other manipulable objects) have specific 'hard-wiring', in that they evoke motor responses even when they are only visually perceived. However, it is also apparent that the management and control of a motor response is covered by an appropriate motor schema, and that the interpretation of the object is covered by appropriate cognitive schema. It is possible that one reason why insects, for instance, *must* respond to an object as a tool in a given context is that they lack the motor and cognitive schema to allow flexible response. In other words, the 'hard-wiring' of visual perception of the tool to the motor cortex (or equivalent structure) means that a response will always be elicited. As I said in Chapter 11, perhaps a defining feature of tool use is the (paradoxical) fact that one can opt to *not* use a given tool in a given situation.

The notion of morphological and motor engagement have already been alluded to in the discussion of motor schema. In broad terms, I am assuming that there is both a 'coordinative' structure, through which well-practised tool-using actions are recorded, and a higher-order control which monitors, corrects and selects from the sets of coordinative structures. Thus, the novice performer will have little option but to use a single coordinative structure, but the expert will have developed a set of such structures and be able to move between them. An useful analogy of this would be to compare the manner in which two people use the bow on a violin: the young child who is learning to play will have a set of actions that concentrate on maintaining contact between bow and string and on moving the bow back and forwards (thus, producing the characteristic squalling sound of the young violinist); the concert violinist, will have a whole host of bowing movements and techniques to produce a wide variety of sounds and effects. In this instance, the tool is similar but its manipulation is significantly changed (indeed, as bowing technique develops, it is likely that the violinist will begin to select bows that are tailored to a particular style of playing).

The notion of perceptual engagement presents an interesting crossover between motor and cognitive schema. In this book, I have been using environmental engagement, in terms of affordances, to describe some of the properties that are usually ascribed to perception (and have referred to the 'direct perception' to describe these). When I talk of perception, I have in mind the interpretation of sensory data. The act of interpretation requires both the prediction or anticipation of data (possibly through motor schema) and the labelling of these data in terms of known meanings (through cognitive schema). Consequently, in tool use perceptual engagement often occurs when the action is not going as planned, i.e., when it is starting to break down. It is the perception of that failing action that leads to decisions to modify, halt or otherwise affect the action.

Cultural engagement is presented in terms of stereotypical responses that people learn in their culture. The culture might be defined as a particular country or might be defined in terms of the working practices of a particular domain. In both cases, there is the 'normal' way of doing things, and this way becomes embedded in our expectations of how things work and the accepted procedures for using these tools. I assume that such knowledge is held primarily in cognitive schema, although, of course, when thinking about population stereotypes and exploratory procedures, it is equally likely that part of this knowledge is held in motor schema.

Over and above all of these different forms of engagement, sits the daemon of cognitive engagement which is both monitoring the moment-by-moment unfolding of an action, preparing and planning subsequent steps and making associations between current state and possible means of reaching a goal state.

Influencing design

Given the notion of forms of engagement, the final part of this book will be an attempt to develop these into design principles. Chapter 6 provided a set of principles for the ergonomics of tools, particularly in terms of design. However, my feeling is that a further set of principles can be developed that are as valid and can be applied to both tools and cognitive artefacts. The basic assumption is that tool use is driven by schema. Consequently, the first set of principles relate to developing designs that encourage particular schema.

1 Use the tool to define its use. This might seem a pretty obvious statement; if a knife has a handle, you hold it. However, a less obvious point is that the weight of a knife, its movement in the hand, the way that the hand can wield the knife and the angle at which the knife can be held for optimal performance are all implicit in the design and tend to only arise when the user has experience in using it. As we saw in Chapter 5, when Seymour was able to define a simple 'vocabulary' of knife grasps and actions, it made training in the use of knives much easier. My proposal is that a well-designed tool, will not only provide the user with somewhere to hold but will also provide indications of points of balance, angle of operation, etc. This does not mean that I am advocating those awful indented handles (I have said in several places that such designs are most assuredly *not* ergonomic). Rather I am proposing that the designers provide a broader set of cues for prospective users. An interesting point from this is how can the design of tool indicate that the user is allowed to employ the tool in a variety of ways?

2 Use the tool to constrain its use. As mentioned in Chapter 6, many jobs require purpose-made tools. This, to me, seems to be an ideal state of affairs to which tool design can aspire. If the tool is made to perfectly match the demands of the task, the characteristics of the user and the environment in which it is to be used, then one can significantly enhance performance. Of course, this is not a cost-effective solution for mass produced tools. However, it strikes me that, like all ideal states, there is no harm in seeking to modify tools in order to strive towards the ideal. From an ergonomics perspective, it might be relatively easy and cheap to change the handles on tools in a particular factory or to make modifications in the workspace to enable better use of the available tools.

3 Design the tool with the workspace. The manner in which a tool is used is influenced by the environment in which it is used. Consequently, the surrounding equipment and the space in which a person works will

have a bearing on how the tool is used. Furthermore, the manner in which it is performed will influence the use of tools. Recall the examples of using power tools in Chapter 8. Many of the problems associated with these tools arose not only from the manner in which the action stopped, but also in terms of the postures that the workplace demanded on the user.

Discussion

In our everyday lives, we are all experts at using the tools that we typically encounter. This 'expertise' is taken for granted and seldom reflected upon. Indeed, the very question of how people perform the mundane and trivial acts involving tools in their everyday lives was the spur to writing this book. Having said this, the ability to wield, manipulate and coordinate tools to perform all manner of tasks is a defining characteristic of human beings. Despite the observations that animals can and sometimes do use tools, there is little evidence to suggest that their lives are as inextricably intertwined with the use of tools as humans. Think for a moment of the physical objects that you use for personal grooming, for eating, for leisure and for work on a daily basis – now imagine not having access to these things.

My aim in writing this book has been to address what I consider to be a fundamental and quotidian aspect of human behaviour: the use of tools. My thesis is that cognition plays a variety of roles in the use of tools, although the more practiced the activity, the more cognition operates on a global rather than local scale.

References

1 Introduction

1 Franklin, 1778, declared that 'Man is a tool using animal', an aphorism to which Carlyle, 1863, added 'without tools he is nothing; with them he is all'.
2 McCullough, M., 1999, *Abstracting Craft*, Cambridge, MA: MIT Press.
3 Vygotsky, L.S., 1928, The instrumental method in psychology, in R.W. Rieber and J. Wollock (eds), 1997, *The Collected Works of L.S. Vygotsky: Volume 3 Problems of the Theory and History of Psychology*, New York: Plenum Press.
4 Butler, S., 1912, On tools, in G. Keynes and B. Hill (eds), 1951, *Samuel Butler's Notebooks*, London: Jonathan Cape, p. 121.
5 Heidegger, M., 1927, *Being and Time*, San Francisco, CA: Harper.
6 Butler, ibid., p. 122.
7 Goodall, J. van Lewick, 1970, Tool using in primates and other vertebrates, in D. Lehrman, R. Hinde and E. Shaw (eds), *Advances in the Study of Behavior Volume 3*, New York: Academic Press, pp. 195–249.
8 Beck, B.B., 1980, *Animal Tool Use Behavior: The Use and Manufacture of Tools by Animals*, New York: Garland STPM Press, p. 122.
9 Beck, ibid., p. 133.
10 Drillis, R.W., 1963, Folk norms and biomechanics, *Human Factors 5*, 427–441.
11 Engelbart, D.C., 1963, A conceptual framework for the augmentation of man's intellect, in Howerton and Weeks (eds), *Vistas in Information Handling Volume 1*, New York: Spartan Books, pp. 1–29.
12 Lave, J., 1988, *Cognition in Practice*, Cambridge: Cambridge University Press.

2 How animals use tools

1 Bowman, R., 1961, Morphological differentiation and adaptation in the Galapagos Finches, *University of California Publications on Zoology 58*, 1–326.
2 Allcock, J., 1972, The evolution of the use of tools by feeding animals, *Evolution 26*, 464–473.
3 Alcock, ibid., p. 472.
4 Beck, 1980, *Animal Tool Use Behavior: The Use and Manufacture of Tools by Animals*, New York: Garland STPM Press.
5 Parker, S. and Gibson, K., 1977, Object manipulation, tool use and sensorimotor intelligence as feeding adaptations in cebus monkeys and great apes, *Journal of Human Evolution 6*, 623–641.

6 Beck, ibid.
7 Alcock, ibid.
8 Miller, G.A., Galanter, E. and Pribram, K.H., 1960, *Plans and the Structure of Behavior*, London: Holt, Rinehart and Winston.
9 Hunt, G., Corballis, M.C. and Gray, R.D., 2001, Laterality in tool manufacture by crows, *Nature 414*, 707.
10 Goodall, J. van-Lewick, 1970, Tool using in primates and other vertebrates, in D. Lehrman, R. Hinde and E. Shaw (eds), *Advances in the Study of Behavior Volume 3*, New York: Academic Press, pp. 195–249.

3 Tool use by primates and young children

1 Goodall, J. van-Lewick, 1968, The behaviour of free-living chimpanzees in the Gombe Stream Reserve, *Animal Behaviour Monographs 1*, 161–311.
2 McGrew, M.C., 1974, Tool use by wild chimpanzees in feeding upon driver ants, *Journal of Human Evolution 3*, 501–508.
3 Nishida, T., 1973, The ant-gathering behavior by the use of tools among wild chimpanzees of the Mahali mountains, *Primates 20*, 357–370.
4 Brewer, S.M. and McGraw, W.C., 1990, Chimpanzee use of a tool-set to get honey, *Folia Primatologica 54*, 100–104.
5 Boesch, C., 1993, Aspect of transmission of tool-use in wild chimpanzees, in K.R. Gibson and T. Ingold (eds), *Tools, Language and Cognition in Human Evolution*, Cambridge: Cambridge University Press, pp. 171–183.
6 Vygotsky, L.S., 1930, Foreword to Köhler's investigation of the intellect of anthropoids, in R.W. Rieber and J. Wollock (eds), 1997, *The Collected Works of L.S. Vygotsky: Volume 3 Problems of the Theory and History of Psychology*, New York: Plenum Press, p. 192.
7 Tanner, N.M., 1988, Becoming human, our links with our past, in T. Ingold (ed.), *What is an Animal?* London: Unwin Hyman, pp. 127–140.
8 Beck, 1980, *Animal Tool Use Behavior: The Use and Manufacture of Tools by Animals*, New York: Garland STPM Press.
9 Beck, ibid.
10 Köhler, 1925, *The Mentality of Apes*, London: Kegan Paul, p. 217.
11 Schiller, P.H., 1949, Manipulative patterns in the chimpanzee, in C.H. Schiller (ed.), 1957, *Instinctive Behavior*, New York: International Universities Press.
12 Vygotsky, ibid., p. 186.
13 Vygotsky, ibid., p. 184.
14 Köhler, ibid., p. 36.
15 Papert, S., 1980, *Mindstorms*, New York: Basic Books.
16 Visalberghi, E., 1993, Capuchin monkeys: a window into tool use in apes and humans, in K.R. Gibson and T. Ingold (eds), *Tools, Language and Cognition in Human Evolution*, Cambridge: Cambridge University Press, pp. 138–150.
17 Beck, ibid.
18 Birch, H., 1945, The relation of previous experience to insightful problem solving, *Journal of Comparative Physiological Psychology 38*, 367–383.
19 Slyva, K., Bruner, J. and Genova, P., 1976, The role of play in the problem-solving of children 3–5 years old, in J. Bruner, A. Jolly and K. Sylva (eds), *Play*, New York: Basic Books, pp. 244–257.

20 Tomasello, M., 1994, The question of chimpanzee culture, in R. Wrangham, W. McGrew, F. de Waal and P. Heltne (eds), *Chimpanzee Cultures*, Harvard: Harvard University Press, p. 305.

4 The making of tools

1 Napier, J., 1980, *Hands*, New York: Pantheon, p. 115.
2 Napier, ibid., p. 115.
3 Gibson, J.J., 1996, *The Senses Considered as Perceptual Systems*, Boston: Houghton Mifflin.
4 Rosenbaum, D.A., 1991, *Human Motor Control*, San Diego, CA: Academic Press, p. 11.
5 Napier, ibid., p. 115.
6 Oakley, 1972, *Man the Tool Maker*, London: The Trustees of the British Museum, p. 3.
7 Leakey, M.D., 1971, *Olduvai Gorge: Excavation in Beds I and II, 1960–1963*, Cambridge: Cambridge University Press.
8 Toth, N. and Schick, K., 1993, Early stone industries and inferences regarding language and cognition, in K.R. Gibson and T. Ingold (eds), *Tools, Language and Cognition in Human Evolution*, Cambridge: Cambridge University Press, p. 349.
9 Oakley, ibid.
10 Toth, N. and Schick, K., 1993, Early stone industries and inferences regarding language and cognition, in K.R. Gibson and T. Ingold (eds), *Tools, Language and Cognition in Human Evolution*, Cambridge: Cambridge University Press, p. 351.
11 Wright, R.V.S., 1972, Imitative learning of a flaked stone technology – the case of an orangutan, *Mankind 8*, 296–306.
12 Savage-Rumbaugh, S. and Lewin, R., 1994, *Kanzi: The Ape at the Brink of the Human Mind*, New York: John Wiley & Sons.
13 Schick, K. and Toth, N., 1993, *Making Silent Stones Speak*, London: Weidenfeld and Nicolson, p. 129.
14 Wynn, T. and Mcgrew, D., 1989, An ape's view of the Oldowan, *Man*, p. 387.
15 Oakley, ibid., p. 44.
16 Oakley, ibid., p. 81.

5 Working with tools

1 Pye, D., 1968, *The Nature and Art of Workmanship*, Cambridge: Cambridge University Press, p. 20.
2 Seymour, W.D., 1966, *Industrial Skills*, London: Pitman, pp. 35–36.
3 Brown, J.S., Collins, A. and Duguid, P., 1899, Situated cognition and the culture of learning, *Educational Researcher 18*, 32–41.
4 Culter, A., 1994, *The Hand of the Master: Craftsmanship, Ivory and Society in Byzantium*, Princeton, NJ: Princeton University Press, p. 80.
5 Veblen, T., 1914, *The Instinct of Workmanship and the State of the Industrial Arts*, New York: MacMillan, p. 304.
6 Chandler, D., 1995, *The Act of Writing*, Aberystwyth: University of Wales.
7 Jean-Paul Sartre, quoted in Chandler, pp. 172–173.

8 McCullough, M., 1996, *Abstracting Craft: The Practiced Digital Hand*, Cambridge, MA: MIT Press, p. 61.

9 Kao, H., 1976, An analysis of user preference toward handwriting instruments, *Perceptual and Motor Skills 43*, 522.

10 Roux, V., Bril, B. and Dietrich, G., 1995, Skills and learning difficulties involved in stone-bead knapping in Khabhat, India, *World Archaeology 37*, 63–87.

11 Suchman, L.A., 1990, *Plans and Situated Actions*, Cambridge: Cambridge University Press.

12 Keller, C.M. and Keller, J. Dixon, 1996, *Cognition and Tool Use: The Blacksmith at Work*, Cambridge: Cambridge University Press.

13 Keller and Keller, ibid., p. 55.

6 The design of tools

1 Drillis, R.W., 1963, Folk norms and biomechanics, *Human Factors 5*, 427–441.

2 Widule, C.J., Foley, V. and Demo, F., 1978, Dynamics of the axe swing, *Ergonomics 21*, 925–930.

3 Greenberg, L. and Chaffin, D., 1977, *Workers and Their Tools*, Midland, MI: Pendell Publishing.

4 Pheasant, S., 1989, *BodySpace*, London: Taylor and Francis.

5 Pheasant, ibid.

6 Drillis, ibid.

7 Freivalds, A., 1987, The ergonomics of tools, in D.J. Oborne (ed.), *International Review of Ergonomics Volume 1*, London: Taylor and Francis, pp. 43–76.

8 Knowlton, R. and Gilbert, J., 1983, Ulnar deviation and short-term strength reduction as affected by a curve-handled ripping hammer and a conventional claw hammer, *Ergonomics 26*, 173–179.

9 Napier, J., 1980, *Hands*, New York: Pantheon.

10 Kroemer, K.H.E., 1986, Coupling the hand with the handle: an improved notation of touch, grip and grasp, *Human Factors 28*, 337–339.

11 Kadefors, R., Areskoug, A., Dahlman, S., Kilbom, A., Sperling, L., Wikström, L. and Oder, J., 1993, An approach to ergonomic evaluation of hand tools, *Applied Ergonomics 24*, 203–211.

12 Drillis, ibid.

13 Greenberg and Chaffin, ibid.

14 Mital, A., 1991, Hand tools: injuries, illnesses, design and usage, in A. Mital and W. Karwowski (eds), *Workspace, Equipment and Tool Design*, Amsterdam: Elsevier, pp. 219–256.

15 Freivalds, A., ibid.

7 The semantics of tools

1 Levi-Strauss, C., 1966, *The Savage Mind*, Chicago: Chicago University Press.

8 How tool use breaks down

1 Reason, J., 1990, *Human Error*, Cambridge: Cambridge University Press.

2 Colley, A.M., 1980, Cognitive motor skills, in D. Holding (ed.), *Human Skills*, Chichester: Wiley, pp. 229–248.

3 http://www.nohsc.gov.au/OHSInformation/Databases/OHSLITPGM/OHSLI/.
4 HSE, 1993, Case Study 4: North Sea Oil Production, *The Costs of Accidents at Work*, London: HMSO.
5 Becker, T.M., Trinkaus, K.M. and Buckely, D.I., 1996, Tool-related injuries among amateur and professional woodworkers, *Journal of Occupational and Environmental Medicine 38*, 1032–1035.
6 HASS, 1999, *Home Accident Surveillance Statistics 1999*, London: Her Majesty's Stationery Office.
7 Aghazadeh, F. and Mital, A., 1987, Injuries due to handtools, *Applied Ergonomics 18*, 273–278.
8 http://www.eig.com/smos/smo97013.html.
9 Gorsche, R., Wiley, J.P., Renger, R., Brant, R., Gemer, T.Y. and Sasyniuk, T.M., 1998, Prevalence and incidence of stenosing flexor tenosynovitis (trigger finger) in a meat-packing plant, *Journal of Occupational and Environmental Medicine 40*, 556–560.
10 HSE, 1990, *Work Related Upper Limb Disorders: A Guide to Prevention*, London: HMSO.
11 Poizner, H., Clark, M.A., Merians, A.S., Macauley, B., Rothi, L.J.G. and Heilman, K.M., 1995, Joint coordination deficits in limb apraxia, *Brain 118*, 227–239.
12 Kempler, D., 1997, Disorders of language and tool use, in K.R. Gibson and T. Ingold (eds), *Tools, Language and Cognition in Human Evolution*, Cambridge: Cambridge University Press, pp. 193–215.
13 Sirigu, A., Cohen, L., Duhamel, J.R., Pillon, B., Dubois, B. and Agid, Y., 1995, A selective impairment of hand posture for object utilization in apraxia, *Cortex 31*, 41–56.
14 Heilman, K.M., Rothi, L.J. and Valenstein, E., 1982, Two forms of ideomotor apraxia, *Neurology 32*, 342–346.
15 Luria, A.R., 1973, The working brain: an introduction to neuropsychology, Harmondsworth: Penguin, pp. 199–200.
16 Roy, E.A., Brown, L. and Hardie, M., 1993, Movement variability in limb gesturing: implications for understanding apraxia, in K.M. Newell and D.M. Corcos (eds), *Variability and Motor Control*, Champaign, IL: Human Kinetics Publishers, pp. 449–473.
17 Rumiati, R.I., Zanini, S., Vorano, L. and Shallice, T., 2001, A form of ideational apraxia as a defective deficit of contention scheduling, *Cognitive Neuropsychology 18*, 617–642.
18 Murata, A., Fadiga, L., Fogassi, L., Gallesse, V., Raos, V. and Rizzolatti, G., 1997, Object representation in the ventral premotor cortex (area F5) of the monkey, *J. Neurophysiol. 78*, 2226–2230.
19 Martin, A., Wiggs, C.L., Ungerleider, L.G. and Haxby, J.V., 1996, Neural correlates of category-specific knowledge, *Nature 379*, 649–652.
20 Chao, L.L. and Martin, A., 2000, Representation of manipulable man-made objects in the Dorsal stream, *NeuroImage 12*, 484–487.
21 Jeannerod, M., 1997, *The Cognitive Neuroscience of Action*, Oxford: Blackwell.
22 Jeannerod, ibid., pp. 92–93.
23 Forde, E. and Humphreys, G.W., 2000, The role of semantic knowledge and working memory in everyday tasks, *Brain and Cognition 44*, 214–252.

9 Cognitive artefacts

1 Vygotsky, L.S., 1928, The instrumental method in psychology, in R.W. Rieber and J. Wollock (eds), 1997, *The Collected Works of L.S. Vygotsky: Volume 3 Problems of the Theory and History of Psychology*, New York: Plenum Press, p. 85.

2 Vygotsky, L., 1978, *Mind in Society: The Development of Higher Psychological Processes*, Cambridge, MA: Harvard University Press, p. 52.

3 Hutchins, E., 1990, The technology of team navigation, in J. Galegher, R.E. Kraut and C. Egido (eds), *Intellectual Teamwork: Social and Technological Foundations of Cooperative Work*, Hillsdale, NJ: LEA, pp. 191–220.

4 Norman, D., 1991, Cognitive artifacts, in J.A. Carroll (ed.), *Designing Interaction*, Cambridge: Cambridge University Press, pp. 17–38.

5 Baber, C., Arvanitis, T.N., Haniff, D.J. and Buckley, R., 1999, A wearable computer for paramedics: studies in model-based, user-centred and industrial design, in M.A. Sasse and C. Johnson (eds), *Interact'99*, Amsterdam: IOS Press, pp. 126–132.

6 Bødker, S., 1990, *Through the Interface – A Human Activity Approach to User Interface Design*, Hillsdale, NJ: LEA.

7 Bartlett, F., 1943, Fatigue following highly skilled work, *Proceedings of the Royal Society, B. vol. 131*, 248–257.

10 Tools in the twenty-first century

1 Braverman, H., 1974, *Labor and Monopoly Capital*, New York: Monthly Review Press.

2 Endsley, M. and Kiris, E.O., 1995, The out-of-the-loop performance problem and level of control in automation, *Human Factors 37*, 381–384.

3 Taylor, R.G., 1978, The metal working machine tool operator, in W.T. Singleton (ed.), *The Analysis of Practical Skills*, Lancaster: MTP Press, pp. 85–111.

4 Coolley, M.J.E., 1988, *Architect or Bee?* London: Hogarth Press.

5 Ulich, E., Schupbach, H., Schilling, A. and Kuark, J.K., 1990, Concepts and Procedures of Work Psychology for the Analysis, Evaluation and Design of Advanced Manufacturing Systems: A Case Study, *International Journal Industrial Ergonomics, 5*, 47–57.

6 Keller, gMBH, CNC+.

7 Buxton, W., 1986, There's more to interaction than meets the eye: some issues in manual input, in D.A. Norman and S.W. Draper (eds), *User Centred System Design*, Hillsdale, NJ: LEA, pp. 319–338.

8 Akamatsu, M. and Sato, S., 1994, A Multi-Modal Mouse with Tactile and Force Feedback, *International Journal of Human-Computer Studies, 40*, 443–453.

9 Gescheider, G.A., Caparo, A.J., Frisina, R.D., Hamer, R.D. and Verillo, R.T., 1978, The effects of a surround on vibrotactile feedback, *Sensory Processes 2*, 99–115.

10 Baber, C., 1997, *Beyond the Desktop*, San Diego, CA: Academic Press.

11 Moody, L., Baber, C. and Arvanitis, T.N., 2002, Objective surgical performance evaluation based on haptic feedback, in J.D. Westood *et al.* (eds), *Medicine Meets Virtual Reality 02/10, Digital Upgrades: Applying Moore's Law, Studies in Health Technology and Informatics*, Amsterdam: IOS Press, pp. 304–310.

12 Wellner, P., 1993, Interaction with paper on the digital desk, *Communications of the ACM 36*, 87–96.

13 Ishii, H. and Ullmer, B., 1997, Tangible bits: towards seamless interfaces between people, bits and atoms, *CHI'97*, New York: ACM, 234–241.

14 Fjeld, M., Voorhost, F., Bischel, M., Lauche, K., Rauterberg, M. and Krueger, H., 1999, Exploring brick-based navigation and composition in an augmented reality, in Gellerson, H-W. (ed.), *Handheld and Ubiquitous Computing*, Berlin: Spriner-Verlag, pp. 102–116.

15 Cooper, L., Johnson, G.I. and Baber, C., 1999, A Run on Sterling – Personal Finance on the Move, *Proceedings of the 3rd International Symposium on Wearable Computers*, Los Alamitos, CA: IEEE Computer Society, 87–92.

16 Gellerson, H-W., Beigl, M. and Krull, H., 1999, Mediacup: awareness technology enabled in an everyday object, in Gellerson, H-W. (ed.), *Handheld and Ubiquitous Computing*, Berlin: Spriner-Verlag, pp. 308–310.

17 Poupyrev, I., Tan, D., Billinghurst, M., Kato, H., Regenbrecht, H. and Tetsutani, N., 2001, Tiles: a mixed reality authoring interface, in Hirose, M. (ed.), Interact'01, Amsterdam: IOS Press, pp. 334–341.

18 Ullmer, B., Ishii, H. and Glas, D., 1998, MediaBlocks: physical containers, transports and controls for online media, *SIGGRAPH'98*, New York: ACM, 379–386.

19 Boud, A., Baber, C. and Steiner, S., 2000, Virtual reality for assembly, *Presence: Teleoperators and Virtual Environments 9 (5)*, 486–496.

20 Baber, ibid., p. 276.

11 Towards a theory of tool use

1 Norman, D.A., 1988, *The Psychology of Everyday Things*, New York: Basic Books, p. 119.

2 Goodale, M.A. and Milner, A.D., 1992, Separate visual pathway for perception and action, *Trends in Neuroscience 15*, 20–25.

3 Gibson, J.J., 1996, *The Senses Considered as Perceptual Systems*, Boston: Houghton Mifflin.

4 Gibson, ibid., p. 46.

5 Lederman, S.J. and Klatzky, R.L., 1996, Action for perception: manual exploratory movements for haptically processing objects and their features, in A.M. Wing, P. Haggard and J.R. Flanagan (eds), *Hand and Brain: The Neurophysiology and Psychology of Hand Movements*, San Diego, CA: Academic Press, pp. 431–446.

6 Jeannerod, M., 1997, *The Cognitive Neuroscience of Action*, Oxford: Blackwell.

7 Jeannerod, ibid.

8 Johansson, R.S. and Westling, G., 1984, Roles of glabrous skin receptors and sensorimotor memory in automatic control of precision grip when lifting rougher or more slippery objects, *Experimental Brain Research 56*, 550–564.

9 Turrell, Y.N., Li, F-X. and Wing, A.M., 1999, Grip force dynamics in the approach to a collision, *Experimental Brain Research 128*, 86–91.

10 Bingham, G.P., 1988, Task-specific devices and the perceptual bottleneck, *Human Movement Science 7*, 225–264.

11 Beek, P.J. and Bingham, G., 1991, Task-specific dynamics and the study of perception and action: a reaction to von Hofsten (1989), *Ecological Psychology 3 [35–54]*, 38–39.

12 Bernstein, N., 1967, *The Coordination and Regulation of Movements*, Oxford: Pergamon Press.

13 Bartlett, F.C., 1932, *Remembering: A Study in Experimental and Social Psychology*, Cambridge: Cambridge University Press.
14 Newell, K.M., 1985, Coordination and control in skill, in D. Goodman, R.B. Wilberg and I.M. Franks (eds), *Differing Perspectives in Motor Learning, Memory and Control*, Amsterdam: Elsevier.
15 Bartlett, ibid.
16 Bartlett, ibid.
17 Bartlett, ibid.
18 Norman, D.A. and Shallice, T., 1980, *Attention to Action: Willed and Automatic Control of Behavior (CHIP Report 99)*, San Diego, CA: University of California.
19 Cooper, R. and Shallice, T., 1997, Modelling the selection of routine actions: exploring the criticality of parameter values, in M.G. Shafto and P. Langley (eds), *Proceedings of the 19th Annual Conference of the Cognitive Science Society*, Stanford, CA: Norton, pp. 131–136.
20 Baber, C. and Stanton, N.A., 1998, Rewritable routines in human interaction with public technology, *International Journal of Cognitive Ergonomics 1 (4)*, 337–349.
21 Keller, C.M. and Keller, J. Dixon, 1996, *Cognition and Tool Use: The Blacksmith at Work*, Cambridge: Cambridge University Press.
22 Turner, P. and Turner, S., 2002, An affordance-based framework for CVE evaluation, in X. Faulkner, J. Finlay and F. Détienne (eds), *People and Computers XVI*, Berlin: Springer, pp. 89–103.

12 Conclusions

1 Guiard, Y., 1987, Assymetric division of labour in human skilled bimanual action: the kinematic chain as a model, *Journal of Motor Behaviour 19*, 486–517.
2 Moody, C.L., Baber, C., Arvanitis, T.N. and Elliott, M., 2003, Objective metrics for the evaluation of simple surgical skills in real and virtual domains, *Presence: Teleoperators and Virtual Environments 12 (2)*.
3 Rasmussen, J., 1979, Outline of a hybrid model of the process plant operator, in T.B. Sheridan and G. Johannsen (eds), *Monitoring Behaviour and Supervisory Control*, New York: Plenum Press, pp. 371–384.

Bibliography

Aghazadeh, F. and Mital, A., 1987, Injuries due to handtools, *Applied Ergonomics* 18, 273–278.

Akamatsu, M. and Sato, S., 1994, A multi-modal mouse with tactile and force feedback, *International Journal of Human-Computer Studies 40*, 443–453.

Allcock, J., 1972, The evolution of the use of tools by feeding animals, *Evolution 26*, 464–473.

Baber, C. and Stanton, N.A., 1998, Rewritable routines in human interaction with public technology, *International Journal of Cognitive Ergonomics 1 (4)*, 337–349.

Baber, C., 1997, *Beyond the Desktop: Designing and Using Interaction Devices*, San Diego, CA: Academic Press.

Baber, C., Arvanitis, T.N., Haniff, D.J. and Buckley, R., 1999, A wearable computer for paramedics: studies in model-based, user-centred and industrial design, in M.A. Sasse and C. Johnson (eds), *Interact '99*, Amsterdam: IOS Press, 126–132.

Bartlett, F., 1943, Fatigue following highly skilled work, *Proceedings of the Royal Society, B. Volume 131*, 248–257.

Bartlett, F.C., 1932, *Remembering: A Study in Experimental and Social Psychology*, Cambridge: Cambridge University Press.

Beck, 1980, *Animal Tool Use Behavior: The Use and Manufacture of Tools by Animals*, New York: Garland STPM Press.

Becker, T.M., Trinkaus, K.M. and Buckely, D.I., 1996, Tool-related injuries among amateur and professional woodworkers, *Journal of Occupational and Environmental Medicine 38*, 1032–1035.

Beek, P.J. and Bingham, G., 1991, Task-specific dynamics and the study of perception and action: a reaction to von Hofsten (1989), *Ecological Psychology 3*, 35–54.

Bernstein, N., 1967, *The Coordination and Regulation of Movements*, Oxford: Pergamon Press.

Bingham, G.P., 1988, Task-specific devices and the perceptual bottleneck, *Human Movement Science 7*, 225–264.

Birch, H., 1945, The relation of previous experience to insightful problem solving, *Journal of Comparative Physiological Psychology 38*, 367–383.

Bødker, S., 1990, *Through the Interface – A Human Activity Approach to User Interface Design*, Hillsdale, NJ: LEA.

Boesch, C., 1993, Aspect of transmission of tool-use in wild chimpanzees, in K.R. Gibson and T. Ingold (eds), *Tools, Language and Cognition in Human Evolution*, Cambridge: Cambridge University Press, 171–183.

Boud, A., Baber, C. and Steiner, S., 2000, Virtual reality for assembly, *Presence: Teleoperators and Virtual Environments 9 (5)*, 486–496.

Bowman, R., 1961, Morphological differentiation and adaptation in the Galapagos Finches, *University of California Publications on Zoology 58*, 1–326.

Braverman, H., 1974, *Labor and Monopoly Capital*, New York: Monthly Review Press.

Brewer, S.M. and McGraw, W.C., 1990, Chimpanzee use of a tool-set to get honey, *Folia Primatologica 54*, 100–104.

Brown, J.S., Collins, A. and Duguid, P., 1899, Situated cognition and the culture of learning, *Educational Researcher 18*, 32–41.

Butler, S., 1912, On tools, in G. Keynes and B. Hill (eds), 1951, *Samuel Butler's Notebooks*, London: Jonathan Cape.

Buxton, W., 1986, There's more to interaction than meets the eye: some issues in manual input, in D.A. Norman and S.W. Draper (eds), *User Centred System Design*, Hillsdale, NJ: LEA, 319–338.

Chandler, D., 1995, *The Act of Writing*, Aberystwyth: University of Wales.

Chao, L.L. and Martin, A., 2000, Representation of manipulable man-made objects in the Dorsal stream, *NeuroImage 12*, 487–484.

Colley, A.M., 1980, Cognitive motor skills, in D. Holding (ed.), *Human Skills*, Chichester: Wiley, 229–248.

Coolley, M.J.E., 1988, *Architect or Bee?* London: Hogarth Press.

Cooper, L., Johnson, G.I. and Baber, C., 1999, A run on sterling – personal finance on the move, *Proceedings of the 3rd International Symposium on Wearable Computers*, Los Alamitos, CA: IEEE Computer Society, 87–92.

Cooper, R. and Shallice, T., 1997, Modelling the selection of routine actions: exploring the criticality of parameter values, in M.G. Shafto and P. Langley (eds), *Proceedings of the 19th Annual Conference of the Cognitive Science Society*, Stanford, CA: Norton, 131–136.

Culter, A., 1994, *The Hand of the Master: Craftsmanship, Ivory and Society in Byzantium*, Princeton, NJ: Princeton University Press.

Drillis, R.W., 1963, Folk norms and biomechanics, *Human Factors 5*, 427–441.

Endsley, M. and Kiris, E.O., 1995, The out-of-the-loop performance problem and level of control in automation, *Human Factors 37*, 381–394.

Engelbart, D.C., 1963, A conceptual framework for the augmentation of man's intellect, in Howerton and Weeks (eds), *Vistas in Information Handling Volume 1*, New York: Spartan Books, 1–29.

Fjeld, M., Voorhost, F., Bischel, M., Lauche, K., Rauterberg, M. and Krueger, H., 1999, Exploring brick-based navigation and composition in an augmented reality, in Gellerson, H-W. (ed.), *Handheld and Ubiquitous Computing*, Berlin: Spriner-Verlag, pp. 102–116.

Forde, E. and Humphreys, G.W., 2000, The role of semantic knowledge and working memory in everyday tasks, *Brain and Cognition 44*, 214–252.

Freivalds, A., 1987, The ergonomics of tools, in D.J. Oborne (ed.), *International Review of Ergonomics Volume 1*, London: Taylor and Francis, 43–76.

Gellerson, H-W., Beigl, M. and Krull, H., 1999, Mediacup: awareness technology enabled in an everyday object, in Gellerson, H-W. (ed.), *Handheld and Ubiquitous Computing*, Berlin: Spriner-Verlag, pp. 308–310.

Gescheider, G.A., Caparo, A.J., Frisina, R.D., Hamer, R.D. and Verillo, R.T., 1978, The effects of a surround on vibrotactile feedback, *Sensory Processes 2*, 99–115.

Gibson, J.J., 1996, *The Senses Considered as Perceptual Systems*, Boston: Houghton Mifflin.

Goodale, M.A. and Milner, A.D., 1992, Separate visual pathway for perception and action, *Trends in Neuroscience 15*, 20–25.

Goodall, J. van Lewick, 1970, Tool using in primates and other vertebrates, in D. Lehrman, R. Hinde and E. Shaw (eds), *Advances in the Study of Behavior Volume 3*, New York: Academic Press, 195–249.

Goodall, J. van-Lewick, 1968, The behaviour of free-living chimpanzees in the Gombe Stream Reserve, *Animal Behaviour Monographs 1*, 161–311.

Gorsche, R., Wiley, J.P., Renger, R., Brant, R., Gemer, T.Y., Sasyniuk, T.M., 1998, Prevalence and incidence of stenosing flexor tenosynovitis (trigger finger) in a meat-packing plant, *Journal of Occupational and Environmental Medicine 40*, 556–560.

Greenberg, L. and Chaffin, D., 1977, *Workers and Their Tools*, Midland, MI: Pendell Publishing.

Guiard, Y., 1987, Assymetric division of labour in human skilled bimanual action: the kinematic chain as a model, *Journal of Motor Behaviour 19*, 486–517.

HASS, 1999, *Home Accident Surveillance Statistics 1999*, London: Her Majesty's Stationery Office.

Heidegger, M., 1927, *Being and Time*, San Francisco, CA: Harper.

Heilman, K.M., Rothi, L.J. and Valenstein, E., 1982, Two forms of ideomotor apraxia, *Neurology 32*, 342–346.

HSE, 1990, *Work Related Upper Limb Disorders: A Guide to Prevention*, London: HMSO.

HSE, 1993, Case Study 4: North Sea Oil Production, *The Costs of Accidents at Work*, London: HMSO.

Hunt, G., Corballis, M.C. and Gray, R.D., 2001, Laterality in tool manufacture by crows, *Nature 414*, 707.

Hutchins, E., 1990, The technology of team navigation, in J. Galegher, R.E. Kraut and C. Egido (eds), *Intellectual Teamwork: Social and Technological Foundations of Cooperative Work*, Hillsdale, NJ: LEA, 191–220.

Ishii, H. and Ullmer, B., 1997, Tangible bits: towards seamless interfaces between people, bits and atoms, *CHI'97*, New York: ACM, 234–241.

Jeannerod, M., 1997, *The Cognitive Neuroscience of Action*, Oxford: Blackwell.

Johansson, R.S. and Westling, G., 1984, Roles of glabrous skin receptors and sensorimotor memory in automatic control of precision grip when lifting rougher or more slippery objects, *Experimental Brain Research 56*, 550–564.

Kadefors, R., Areskoug, A., Dahlman, S., Kilbom, A., Sperling, L., Wikström, L. and Oder, J., 1993, An approach to ergonomic evaluation of hand tools, *Applied Ergonomics 24*, 203–211.

Kao, H., 1976, An analysis of user preference toward handwriting instruments, *Perceptual and Motor Skills 43*, 522.

Keller, C.M. and Keller, J. Dixon, 1996, *Cognition and Tool Use: The Blacksmith at Work*, Cambridge: Cambridge University Press.

Kempler, D., 1997, Disorders of language and tool use, in *Tools, Language and Cognition in Human Evolution*, Cambridge: Cambridge University Press, 193–215.

Knowlton, R. and Gilbert, J., 1983, Ulnar deviation and short-term strength reduction as affected by a curve-handled ripping hammer and a conventional claw hammer, *Ergonomics 26*, 173–179.

Köhler, 1925, *The Mentality of Apes*, London: Kegan Paul.

Kroemer, K.H.E., 1986, Coupling the hand with the handle: an improved notation of touch, grip and grasp, *Human Factors 28*, 337–339.

Lave, J., 1988, *Cognition in Practice*, Cambridge: Cambridge University Press.

Leakey, M.D., 1971, *Olduvai Gorge: Excavation in Beds I and II, 1960–1963*, Cambridge: Cambridge University Press.

Lederman, S.J. and Klatzky, R.L., 1996, Action for perception: manual exploratory movements for haptically processing objects and their features, in A.M. Wing, P. Haggard and J.R. Flanagan (eds), *Hand and Brain: The Neurophysiology and Psychology of Hand Movements*, San Diego, CA: Academic Press, 431–446.

Levi-Strauss, C., 1966, *The Savage Mind*, Chicago: Chicago University Press.

Luria, A.R., 1973, *The Working Brain: An Introduction to Neuropsychology*, Harmondsworth: Penguin, 199–200.

Martin, A., Wiggs, C.L., Ungerleider, L.G. and Haxby, J.V., 1996, Neural correlates of category-specific knowledge, *Nature 379*, 649–652.

McCullough, M., 1996, *Abstracting Craft: The Practiced Digital Hand*, Cambridge, MA: MIT Press.

McGrew, M.C., 1974, Tool use by wild chimpanzees in feeding upon driver ants, *Journal of Human Evolution 3*, 501–508.

Miller, G.A., Galanter, E. and Pribram, K.H., 1960, *Plans and the Structure of Behavior*, London: Holt, Rinehart and Winston.

Mital, A., 1991, Hand tools: injuries, illnesses, design and usage, in A. Mital and W. Karwowski (eds), *Workspace, Equipment and Tool Design*, Amsterdam: Elsevier, 219–256.

Moody, C.L., Baber, C., Arvanitis, T.N. and Elliott, M., 2003, Objective metrics for the evaluation of simple surgical skills in real and virtual domains, *Presence: Teleoperators and Virtual Environments 12 (2)*.

Moody, L., Baber, C. and Arvanitis, T.N., 2002, Objective surgical performance evaluation based on haptic feedback, in J.D. Westood *et al.* (eds), *Medicine Meets Virtual Reality 02/10, Digital Upgrades: Applying Moore's Law, Studies in Health Technology and Informatics*, Amsterdam: IOS Press, 304–310.

Murata, A., Fadiga, L., Fogassi, L., Gallesse, V., Raos, V. and Rizzolatti, G., 1997, Object representation in the ventral premotor cortex (area F5) of the monkey, *Journal of Neurophysiology 78*, 2226–2230.

Napier, J., 1980, *Hands*, New York: Pantheon.

Newell, K.M., 1985, Coordination and control in skill, in D. Goodman, R.B. Wilberg and I.M. Franks (eds), *Differing Perspectives in Motor Learning, Memory and Control*, Amsterdam: Elsevier.

Nishida, T., 1973, The ant-gathering behavior by the use of tools among wild chimpanzees of the Mahali mountains, *Primates 20*, 357–370.

Norman, D.A., 1988, *The Psychology of Everyday Things*, New York: Basic Books.

Norman, D., 1991, Cognitive artifacts, in J.A. Carroll (ed.), *Designing Interaction*, Cambridge: Cambridge University Press, 17–38.

Norman, D.A. and Shallice, T., 1980, *Attention to Action: Willed and Automatic Control of Behavior (CHIP Report 99)*, San Diego, CA: University of California.

Oakley, K. 1972, *Man the Tool Maker*, London: The Trustees of the British Museum, p. 3.

Papert, S., 1980, *Mindstorms*, New York: Basic Books.

Parker, S. and Gibson, K., 1977, Object manipulation, tool use and sensorimotor intelligence as feeding adaptations in cebus monkeys and great apes, *Journal of Human Evolution 6*, 623–641.

Pheasant, S., 1989, *BodySpace*, London: Taylor and Francis.

Poizner, H., Clark, M.A., Merians, A.S., Macauley, B., Rothi, L.J.G. and Heilman, K.M., 1995, Joint coordination deficits in limb apraxia, *Brain 118*, 227–239.

Poupyrev, I., Tan, D., Billinghurst, M., Kato, H., Regenbrecht, H. and Tetsutani, N., 2001, Tiles: a mixed reality authoring interface, In Hirose, M. (ed.), Interact'01, Amsterdam: IOS Press, 334–341.

Pye, D., 1968, *The Nature and Art of Workmanship*, Cambridge: Cambridge University Press.

Rasmussen, J., 1979, Outline of a hybrid model of the process plant operator, in T.B. Sheridan and G. Johannsen (eds), *Monitoring Behaviour and Supervisory Control*, New York: Plenum Press, 371–384.

Reason, J., 1990, *Human Error*, Cambridge: Cambridge University Press.

Rosenbaum, D.A., 1991, *Human Motor Control*, San Diego, CA: Academic Press.

Roux, V., Bril, B. and Dietrich, G., 1995, Skills and learning difficulties involved in stone-bead knapping in Khabhat, India, *World Archaeology 37*, 63–87.

Roy, E.A., Brown, L. and Hardie, M., 1993, Movement variability in limb gesturing: implications for understanding apraxia, in K.M. Newell and D.M. Corcos (eds), *Variability and Motor Control*, Champaign, IL: Human Kinetics Publishers, 449–473.

Rumiati, R.I., Zanini, S., Vorano, L. and Shallice, T., 2001, A form of ideational apraxia as a delective deficit of contention scheduling, *Cognitive Neuropsychology 18*, 617–642.

Savage-Rumbaugh, S. and Lewin, R., 1994, *Kanzi: The Ape at the Brink of the Human Mind*, New York: John Wiley & Sons.

Schick, K. and Toth, N., 1993, *Making Silent Stones Speak*, London: Weidenfeld and Nicolson, 129.

Schiller, P.H., 1949, Manipulative patterns in the chimpanzee, in C.H. Schiller (ed.), 1957, *Instinctive Behavior*, New York: International Universities Press.

Seymour, W.D., 1966, *Industrial Skills*, London: Pitman, 35–36.

Sirigu, A., Cohen, L., Duhamel, J.R., Pillon, B., Dubois, B. and Agid, Y., 1995, A selective impairment of hand posture for object utilization in apraxia, *Cortex 31*, 41–56.

Sperling, L., Dahlman, S., Wikström, L., Kilbom, A. and Kadefors, R., 1999, A cube model for the classification of work with hand tools and the formulation of functional requirements, *Applied Ergonomics 24 (3)*, 212–220.

Suchman, L.A., 1990, *Plans and Situated Actions*, Cambridge: Cambridge University Press.

Sylva, K., Bruner, J. and Genova, P., 1976, The role of play in the problem-solving of children 3–5 years old, in J. Bruner, A. Jolly and K. Sylva (eds), *Play*, New York: Basic Books, 244–257.

Tanner, N.M., 1988, Becoming human, our links with our past, in T. Ingold (ed.), *What is an Animal?* London: Unwin Hyman, 127–140.

Taylor, R.G., 1978, The metal working machine tool operator, in W.T. Singleton (ed.), *The Analysis of Practical Skills*, Lancaster: MTP Press, 85–111.

Tomasello, M., 1994, The question of chimpanzee culture, in R. Wrangham, W. McGrew, F. de Waal and P. Heltne (eds), *Chimpanzee Cultures*, Harvard: Harvard University Press, 301–317.

Toth, N. and Schick, K., 1993, Early stone industries and inferences regarding language and cognition, in K.R. Gibson and T. Ingold (eds), *Tools, Language and Cognition in Human Evolution*, Cambridge: Cambridge University Press.

Turner, P. and Turner, S., 2002, An affordance-based framework for CVE evaluation, in X. Faulkner, J. Finlay and F. Détienne (eds), *People and Computers XVI*, Berlin: Springer, 89–103.

Turrell, Y.N., Li, F-X. and Wing, A.M., 1999, Grip force dynamics in the approach to a collision, *Experimental Brain Research 128*, 86–91.

Ulich, E., Schupbach, H., Schilling, A. and Kuark, J.K., 1990, Concepts and procedures of work psychology for the analysis, evaluation and design of advanced manufacturing systems: a case study, *International Journal Industrial Ergonomics 5*, 47–57.

Ullmer, B., Ishii, H. and Glas, D., 1998, MediaBlocks: physical containers, transports and controls for online media, *SIGGRAPH'98*, New York: ACM, 379–386.

Veblen, T., 1914, *The Instinct of Workmanship and the State of the Industrial Arts*, New York: MacMillan.

Visalberghi, E., 1993, Capuchin monkeys: a window into tool use in apes and humans, in K.R. Gibson and T. Ingold (eds), *Tools, Language and Cognition in Human Evolution*, Cambridge: Cambridge University Press 138–150.

Vygotsky, L., 1978, *Mind in Society: The Development of Higher Psychological Processes*, Cambridge, MA: Harvard University Press.

Vygotsky, L.S., 1930, Foreword to Köhler's investigation of the intellect of anthropoids, in R.W. Rieber and J. Wollock (eds), 1997, *The Collected Works of L.S. Vygotsky: Volume 3 Problems of the Theory and History of Psychology*, New York: Plenum Press.

Vygotsky, L.S., 1928, The instrumental method in psychology, in R.W. Rieber and J. Wollock (eds), 1997, *The Collected Works of L.S. Vygotsky: Volume 3 Problems of the Theory and History of Psychology*, New York: Plenum Press.

Wellner, P., 1993, Interaction with paper on the digital desk, *Communications of the ACM 36*, 87–96.

Widule, C.J., Foley, V. and Demo, F., 1978, Dynamics of the axe swing, *Ergonomics 21*, 925–930.

Wright, R.V.S., 1972, Imitative learning of a flaked stone technology – the case of an orangutan, *Mankind 8*, 296–306.

Wynn, T. and Mcgrew, D., 1989, An ape's view of the Oldowan, *Man*, 387.

Name index

Subject index

Milton Keynes UK
Ingram Content Group UK Ltd.
UKHW040054071024
449327UK00019B/557